David Hutchings is a ph[...]
York, England. A Fellow o[...] Institute of Physics, he has written
several books about the relationship between science and religion
and speaks regularly on the topic around the country at conferences,
schools, universities and churches. David has also run multiple
training events for science teachers, specializing in dealing with
common misconceptions in the discipline. He lives in York with his
wife and two young daughters.

David Wilkinson is Principal of St John's College and Professor in
the Department of Theology and Religion at Durham University. He
lives in Newcastle with his wife Alison and has two grown-up children. He is a writer and speaker on Christianity and science not just
in the UK but around the world. He has doctorates in astrophysics
and theology and is a Fellow of the Royal Astronomical Society. He
is also a Methodist minister and the author of many books.

DAVID HUTCHINGS & DAVID WILKINSON

GOD, STEPHEN HAWKING AND THE MULTIVERSE

WHAT HAWKING SAID AND WHY IT MATTERS

First published in Great Britain in 2020

Society for Promoting Christian Knowledge
36 Causton Street
London SW1P 4ST
www.spck.org.uk

British Library Cataloguing-in-Publication Data
A catalogue record for this book is available from the British Library

ISBN 978–0–281–08191–2
eBook ISBN 978–0–281–08192–9

1 3 5 7 9 10 8 6 4 2

Typeset by Manila Typesetting Company
First printed in Great Britain by Jellyfish Print Solutions
Subsequently digitally printed in Great Britain

eBook by Manila Typesetting Company

Produced on paper from sustainable forests

David Hutchings
For Bethany:
my Little Girl who asks Big Questions

David Wilkinson
For John Polkinghorne:
teacher and mentor

Contents

Contents

Contents

Contents

List of plates

Acknowledgements

David Hutchings

When I was still a teenager, somebody (I can't remember who) bought me *God, the Big Bang and Stephen Hawking* by an astrophysicist-theologian named David Wilkinson. This is important for two reasons: first, it makes it clear how much younger I am than David; second, it shows that the God-and-science conversation is one that I have been interested in for a long time. Having the chance to write about it now – and with David, no less – is the realization of a dream that I didn't really dare to have in the first place.

The topics of quantum mechanics, general relativity, cosmological origins and the existence of God are Big. So Big, in fact, that no one can realistically attempt to write about them without a huge amount of help. Many experts have been patient and kind enough to contribute to this project, including Luke Barnes, Geoffrey Cantor, William Lane Craig, Reed Guy, Richard Keesing, Tom Lancaster, Tom McLeish, Matt Probert and Aron Wall. All of these folk are deep thinkers who really know their stuff. Their advice has been invaluable.

Special thanks must also go to our guinea-pig readers: Angie Edwards and Martin Steel. They have pored over every word and set us straight many times on how to go about making some pretty inaccessible ideas a little more accessible. Also qualifying for special thanks is my sixth-form student Magnus Swann. He drew all the figures for this book while simultaneously working for his Physics AS Level. My apologies if this has detracted at all from his result. Tony Collins, our publishing powerhouse, deserves thanks for bringing this book into existence by the sheer force of his personality – he has kept us focused and on task like a proficient classroom teacher. I am also indebted to Michelle Clark and her seemingly infinite capacity to put up with and then sort out other people's mistakes.

Finally, to my family – Emma, Bethany and Chloe – I love you. Thank you for putting up with me 'working on the book'. You are the best. I'm not going to say that I couldn't do it (write books) without you, because I could. It's just that I would be miserable and lonely and probably a horrible person.

David Wilkinson

Compared to David's youthful mind, I am so old that I've probably forgotten all those who have contributed to the ideas and arguments of this book! But I am grateful to Sir Robert Boyd who, long ago, helped me to see the importance of a Christian doctrine of creation in the light of Hawking's work. In addition to the names above, I am personally thankful for study leave from St John's College and, as always, the encouragement and understanding of Alison, Adam and Hannah.

1

Project Shangri-La

The first widely used electronic computers had brains made out of cardboard. Known as punch-cards – stiff pieces of paper with holes in specific locations – they contained all the instructions the computers needed to complete their super-human tasks.

From the 1950s to the 1970s, businesses craved this new-fangled ability to perform important calculations at whirlwind speed – and they were prepared to pay big bucks to the likes of IBM for the privilege of being able to do so.

Big business was not the only computer-consumer; governments saw the advantages too, using computers to track census data and monitor taxes. There can be no doubt about it – automated information processing was the new (and ultra-smart) kid on the block.

The world was hurtling forward faster than ever before and thousands of ingenious new programs were being punched into card on almost a daily basis, each promising its own mini-revolution. Despite the staggering number of instruction-sets being produced, one now-infamous stack of cards can still claim to stand out from all the rest. This was the deck that contained the 'God-Naming Routine' – or *Project Shangri-La*, as those involved had dubbed it.

In 1953, Manhattan computer scientist Dr Julien Wagner struck literal gold when he received a highly unusual request from a group of monks from the wilds of Tibet. Incredibly, this isolated band, hidden away in the perilous heights of the Himalayas, already had their own diesel generator – and they had also (somehow) become aware of Wagner's cutting-edge Automatic Sequence Computer: the Mark V.

In exchange for vast wealth accumulated over centuries (possibly millennia), Wagner's team was asked to write a program that would aid the monks in what they considered to be the most sacred of all human quests: determining the true names of God. This was mankind's sole purpose, they believed, and the order had been studying the matter for at least three hundred years.

While it is tempting to imagine that such a campaign would be deeply mystical in nature – perhaps names were found through meditation or trance or desperate prayer – the monks were instead quite prepared to be pragmatic in their approach. Over the decades, they had systematized the search, steadily working their way through all the possible permutations of letters in their holy alphabet. Their doctrine told them that there were nine billion true names of God and, once they were all found, God's purpose for the universe would be complete.

There was, as it happens, an oddly admirable selflessness about this process: the leader of the monks – who actually made the trip to New York – informed Wagner that if they continued to use the traditional methods, at their current rate, they would require another *fifteen thousand years*.

Amused and fascinated (and presumably also persuaded by the hoard of gold on offer), Wagner's team began adapting the Mark V to the strange alphabet and still-yet-stranger task. Once ready, the programmed computer was capable of printing out every conceivable name that would comply with the monks' mystical rule-set in just 100 days of operation. Wagner sent the machine, along with two technicians, on a plane to India, where they had onward transport arranged by the monks – all expenses paid, of course.

Precisely what happened next is hard to piece together, but not impossible. What is certain is that neither technician returned.

One of them, George Hanley, kept a diary and from this the extraordinary tale of the nine billion names of God can be at least partially deduced.

His colleague on the mission, Finn 'Chuck' Byrne (the origin of the nickname is a mystery), was a very bright chap and had come to the sudden and terrifying realization that the two of them might shortly be in a great deal of trouble.

Here they were, thousands of feet up, in a monastery that might well have been the most remote building on the whole planet, living with several hundred monks who were wholly devoted to just one thing: the names. Indeed, they had spent their entire lives thus far writing them out by hand – it was what they were living for, if it could truly be called living.

What's more, Hanley and Chuck had been treated like royalty. To their surprise and amazement, they had been given Cuban cigars, cashmere coats and – unbelievably – a Jaguar XK120 *each*. How these extravagances had been obtained (and how they were expected to get them home), they had no idea. None of the other monks spoke any English at all and they had barely seen the leader – whom they had begun to call 'Sam' – since arriving and setting the Mark V going.

All the luxury was wonderful at first, but now it had Chuck worried. The monks themselves did not value the cigars, nor the cars. Neither did they value the huge pile of wealth that Wagner had been promised on completion. *Why?*

Taking the opportunity when it came, Chuck had grilled Sam on what was going on. The answer chilled him to the bone. The community did not treasure their treasures because, when the names were all discovered, the universe in its entirety would 'no longer be necessary', which, Wagner's two employees reasoned, supposedly meant the end of the world.

Of course, Chuck and Hanley knew that the world *wouldn't* end along with the program. And, when it didn't, what would happen to them? They didn't much fancy facing a gang of furious – perhaps even murderous – mystics.

After considering sabotaging the computer and then deciding against it, the two men thought the best option was to run for safety – albeit across miles of extreme terrain – without giving prior notice to their soon-to-be-disappointed customers. Gathering what they could, they left the monastery on the very evening that the Mark V would finish.

But Hanley and Chuck didn't make it home. Their bodies were never found; their packed-at-the-last-minute bags were. Hanley's final diary entry is haunting to this day:

Have now left with C
S-La complete soon
Hoping that th
THE STARS ARE GOING OUT

* * *

There is little doubting that this is a gripping story, but what is it about it which unsettles us even now? Surely its power lies in the fact that it draws on so many of the questions fundamental to us – to our very *humanity*. Who are we? What are we here for? What else is out there? Is there a God? Can we know Him/Her/It? What is the universe? Why does it exist? Where did it come from? Does science answer these questions or are they the domain of religion?

There is a twist in the tale of Wagner, his Mark V and *Project Shangri-La*. It is not a story from our universe at all, but from a parallel universe – one found in the mind of the remarkable science fiction author Arthur C. Clarke. The short story of the unfortunate Hanley and Chuck, 'The nine billion names of God', was first published in *Star Science Fiction Stories #1* in 1953 – and has been reproduced many times since.[1] Even now, it remains one of the best examples of the genre.

That does, of course, mean, the events recorded above never really happened. Or, rather, we should maybe say that they *probably* never happened. For if, as some cosmologists have been suggesting for a while, there is actually an infinite number of universes, then they *have* happened – what's more, they have happened an infinite number of times. Poor Chuck.

The crazy consequences of ideas like this explain why the notion of parallel universes – or of a 'multiverse' – is garnering so much attention. What was once the playground of sci-fi writers is now part of the mainstream conversation and speculation is the name of the game. Before we discuss the multiverse and other highly befuddling cosmoddities, however, we must discuss the life story of a man who is so extraordinary that he may as well have come from another universe himself: Professor Stephen Hawking.

The icon

Even Arthur C. Clarke would not have dared to pen a narrative quite like the life and times of Hawking – it would have seemed too much of a stretch. A brief summary makes that point well: born in Oxford in 1942, Stephen Hawking didn't always do all that well at school (coming bottom in his class at times) and yet he ended up with a scholarship to Oxford University no less. Once there, he only worked for about an hour a day, but still finished with a First Class physics degree and gained a graduate place at Cambridge University.

Over the New Year period from 1962 to 1963, he was diagnosed with a fatal degenerative condition known as motor neurone disease and given two years to live. Defying this, he went on to complete his graduate studies, get married and father three children. He also developed radical new theories about time, about mysterious objects known as black holes and even about the origins of the universe itself. Throughout these successes, his physical health continued to deteriorate. Wheelchair-bound and eventually becoming almost totally paralysed, he contracted pneumonia in 1985 – at which point his wife was advised by the doctors to let him die in the hospital.

Once again, though, Hawking beat the odds and survived – this time at the cost of his voice. Despite initial depression, he bounced back, gaining access to a new technology that would speak for him electronically. The strange robotic tones had an accent hovering between American and Scandinavian, yet the unsinkable Hawking came to think of it as 'his' voice; it was to become one of the most recognizable on the planet.

His profile, in and out of academia, grew yet further. He was awarded perhaps the most prestigious scientific Chair (Newton's) in perhaps the most prestigious university in the world (Cambridge); he guested for *Pink Floyd*, *The Simpsons*, *The Big Bang Theory* and *Star Trek*. He became a political and environmental activist, beloved by those with no interest in his field – applied mathematics – whatsoever.

Hawking's esoteric PhD thesis ('Properties of expanding universes', 1966) was made available online in October 2017 – and immediately crashed the University of Cambridge's website. Accessed from every country in the world, it has been viewed millions of times, boasting as many downloads as all the other available documents combined.[2] A Hollywood blockbuster based on his life, *The Theory of Everything*, was nominated for Best Picture at the 2014 Oscars – where the actor playing him, Eddie Redmayne, won Best Actor.

This exceptional man lived to the ripe old age of 76 – an astonishing 2,700 per cent longer than he had been told he would – and was laid to rest in Westminster Abbey between Isaac Newton and Charles

Darwin. The renowned composer Vangelis, of *Chariots of Fire* fame, penned a score for his memorial service, after which arrangements were made to beam some of Hawking's most inspirational thoughts into space, towards a black hole. When Hawking died, after a lifelong story of struggles and victories, he was the most famous scientist in the universe.

Try getting people to swallow all that, Mr Clarke . . .

The book

There is one notable achievement of Hawking's that we have not yet mentioned. In fact, a decent case could be made that, in spite of his undoubted contributions to cosmology, mathematics and entertainment, his greatest feat was one that no one really expected to amount to all that much – a short, and rather odd, *book*.

A Brief History of Time hit the shelves in 1988 and caught everybody out. It would be an exaggeration to say that there was no market at all for popular science books, but when just a few thousand sales could get you on to bestseller lists, expectations are set at that type of level. Sure, there had been some big breakthrough titles, most significantly Carl Sagan's *Cosmos* (1980), but these were often released in tandem with TV programmes and, even then, were a dramatic exception.

That a manuscript, therefore, in which the major themes are gravitationally altered geometries, minority interpretations of quantum mechanics and the importance of multiplying the time dimension by the square root of minus-one might outsell a bodice-ripper or a murder mystery didn't feature on anyone's radar. As Hawking himself says in the introductory material to the 1996 edition of his book:

I don't think anyone, my publishers, my agent, or myself, expected the book to do anything like as well as it did.

It was on the London *Sunday Times* bestseller list for 237 weeks . . .

I have sold more books on physics than Madonna has on sex.[3]

In fact, at the time of writing, *Brief History* has sold over 20 million copies. This is genuinely staggering, yet it is another part of the Hawking story that seems to go entirely against what anyone might usually expect. What's more, he hardly trivialized the subject matter – the reader is not spoken down to and technical topics are not dodged.

Indeed, it has become a truism that members of the public bought the book, placed it on their coffee tables or some other prominent place and never opened it; it is the book that everyone buys and no one reads. When asked about this by the BBC presenter Sue Lawley in 1992,[4] Hawking replied that the scarily bright Bernard Levin – the 'most famous journalist of his day' according to *The Times*[5] – had admitted quitting after just page 29.

This reputation is actually a little undeserved. *Brief History* is not impenetrable; neither is it boring or irrelevant. For its time, it is actually surprisingly accessible and there is a purity about Hawking's refusal to either dress up or apologize for the physics. He was a man fascinated by the Big Questions that could be asked and by the power of maths to find answers. *Brief History* reads like it was written by an enthusiast, in the same way that someone who builds model railways loves to explain to anyone and everyone how they made, from scratch, their very own small-scale trees. As human beings, when we really love something, we want to talk about it with others – and *Brief History* was Hawking's unembarrassed love for mathematical cosmology on full display.

It is true, however, that Hawking was (at best) a minor celebrity at the time of publication – and that the subject matter of the work is undeniably complex. Add the consideration that the average Joe has probably never had a conversation about – or even an interest in – the problems with the infinities thrown up by gravitational singularities in space–time, and the question remains: why did *Brief History* sell so well?

Big Questions

Unsurprisingly, this has been asked plenty of times in the past. Perhaps the people most interested in the answer are publishers: a breakthrough book like *Brief History* is a cash cow and provides opportunities for spin-offs to make even more money. Similarly, science writers dream of hitting those numbers and having their books sitting alongside stories of vampires, wizards and hard-bitten detectives – or weight-loss cookbooks.

It would be naive to think that the success of *Brief History* can really be separated from the person of Hawking himself. His own story is compelling; it is uplifting; it is moving. The man is fascinating. He is 'other' in that his mind is capable of exceptional mathematical insight and his physical condition is so extreme and visible and rare; yet he is 'like us' in that he is so human, so childlike in unabashedly loving his hobby and somehow so utterly authentic.

His sense of humour – exemplified by the Madonna reference mentioned earlier – in the face of such prolonged and lonely suffering shows a bravery that it is impossible not to admire; his wonderful relationship with his daughter Lucy (they have written space stories together) is inspirational; his patience with uncomfortable, insensitive or pitifully predictable interviewers is borderline-unbelievable.

Yet Hawking chooses not to point to any of these things when *he* tries to explain the overwhelming triumph of his first book. Instead, he points to the topics he is writing about and claims that they tap into an aspect of our humanness that we simply cannot leave be: 'The success of *A Brief History* indicates that there is widespread interest in the big questions like: where did we come from? And why is the universe the way it is?'[6]

Surely Hawking is right here – after all, the millions of copies sold were bought by *people*. It would seem, then, that the perfect pairing of Hawking's profound life story and his cutting-edge book about our origins has resulted in something almost primal being triggered within us all: for we *need* to know who we are.

Where do we go from here?

Stephen Hawking wrote many more books after *Brief History* and one of particular note is *The Grand Design*. Co-written with Leonard Mlodinow in 2010, it appeared with the subtitle *New answers to the ultimate questions in life*.

In this present volume, we will be looking at these 'ultimate questions' and the answers Hawking gives to them. To do so, we will need to learn some physics, so the first few chapters will cover the key ideas needed to access Hawking's arguments.

Hawking, like all physicists, builds on the work of those who have gone before him – whether in gravity, quantum theory or cosmology. This means that there are stories to be told about the foundations of physics and these are necessary in order to understand his work. Once those foundations are in place, we will carefully review the theories put forward in *Brief History* and *Grand Design*, analysing the universe that they claim exists.

As we do, it is important to understand that Hawking was a *theoretician* – much of his work was rather abstract and specula-tive, not confirmed yet in any way by experiment. In some cases, it is quite possible that it will never be confirmed. Because of this, we will need to see what other physicists have said about the same topics and the same data – do they agree with Hawking or consider him off-piste? What is the current standing, within the industry, of his ideas?

Once we are finished with the physics, it will be time to get into some philosophy, for the two fields are far more intimately related than many realize. What does Hawking think are the consequences, on the ground, of his theories? Which of these matter to our previously mentioned average Joe? Which of them don't?

From there, we will climb to the pinnacle of question-posing and address the greatest topic of all: God. Hawking and God are, put simply, inseparable. God is named on many of the pages of *Brief History* and *Grand Design*, and indirectly there on those with no mentions of him. Both books deal up-close-and-personal with the God question. It is to Hawking's great credit that he doesn't duck the issue – but does he do it justice? What do the theologians think?

Along the way, we will revel in the bizarreness of black holes; time will disappear before our very eyes; the multiverse will terrify us and render science useless; the existence of reality itself will come under question; God will be off and then on the table before repeating the trick; Hawking will make us laugh and cry with his wit and honesty – and, thankfully, all of this will happen without an equation in sight.

God, Stephen Hawking and the Multiverse is a book that can be read by people who have no idea what the multiverse is, what quantum gravity means or how time could possibly be imaginary. Likewise, it can be read by those who have never studied philosophy, have not thought about the meaning of life or to whom the Bible is a mystery. Yes, it will be a whistle-stop tour through gravitational fields, multiple histories, the New Testament and a little bit of Enlightenment philosophy – but it will be worth it.

Because, as Hawking reminds us in *The Grand Design*:

What is the nature of reality? Where did all this come from? Did the Universe need a Creator? Most of us do not spend most of our time worrying about these questions, but *almost all of us worry about them some of the time.*[7]

2

Tunguska • Elementary, my dear Aristotle • It's all Greek to me • This most beautiful system • Force fields and the art of science • Light and momentary troubles • A space-, time- and mind-bending solution • Hawking, relativity and those Big Questions

Tunguska

In its long-running *Notes & Queries* section, *The Guardian* newspaper 'invites readers to send in questions and answers on everything from trivial flights of fancy to the most profound concepts'. They clearly oblige – for on 14 October 1991, the editors published the following head-scratcher: 'Is it true that a Black Hole hit Siberia early this century?'

As we will discover through the course of this book, there is quite a lot of weird and wonderful physics to be investigated before a detailed answer can be given to this question. A shorter answer, on the other hand, is readily available. It's 'No.'

What is of more interest to us for the moment, however, is why the reader might have asked this question in the first place. For, as it turns out, something rather unusual *did* happen in Siberia early last century. And, though it might well sound like yet another story that has been lifted from the pages of a science fiction novel, this one is *real*.

On the morning of 30 June 1908, somewhere near the Stony Tunguska River in Eastern Siberia, something exploded. Whatever it was, the effect was catastrophic. It flattened nearly a thousand square miles of forest, knocked eighty million trees pancake-flat and sent pressure waves through the air that could be detected as far away as England.

Indeed, so much dust was thrown up into the atmosphere that scattered sunlight lit up the next two *nights* across parts of Europe and Asia. These bizarre circumstances led to reports of people reading newspapers at midnight and even of one chap playing a round of golf at St Andrews – at half past two in the morning.

Due to the complete lack of any local population, no one was killed – but *The Observer* later reported that: 'many Russians, believing this to be a sign of the approaching end of the world, left their homes and belongings, and wandered off to holy shrines and monasteries'.[1]

Modern estimates of the force of the blast put it at somewhere around 15 megatons of TNT, that is one thousand times greater than that of the nuclear bomb dropped on Hiroshima. What could possibly have caused such devastation?

Speculation, as one might expect, has been rife. Study after study has considered the possibilities, each one able to draw on further geographical and theoretical work. To date, more than a thousand academic papers have been published on what has become known as the *Tunguska event* and multiple conspiracies have had their day in the sun along the way. If it was simply a meteor strike, people asked, where was the iron you would expect to find afterwards? More to the point, why wasn't there a crater?

Intriguingly, different eyewitnesses reported a cylinder-shaped object heading in different directions – was this a craft being steered by an alien visitor that then crashed? A (slightly) more serious suggestion was that it may well have been a black hole passing through the Earth. A (slightly) less believable claim was that a Second World War bomber had inadvertently time-travelled back through the Bermuda Triangle before dropping its nuclear payload on to the frozen desert. Unsurprisingly, the allure of the mystery was not lost on the entertainment industry: *The X-Files* and Nintendo have featured Tunguska in TV drama and video game respectively.

So, what *did* happen? Duncan Steel, an astronomer and writer (and good friend of the aforementioned Arthur C. Clarke), pulled together all the available material into one coherent whole for *Nature* on the 100th anniversary of the Tunguska event.[2] Sifting through the vast wealth of data and analysis, he finally arrived at a conclusion – one that (disappointingly) did not involve aliens or time travel or black holes.

Instead, Steel decided, a 50-metre-wide fragment of a comet, travelling at nearly 70,000 mph, had hurtled through our atmosphere and exploded above the surface of the taiga just before impact. Although it was 'only' a fragment, this ball of dusty ice carried more energy than is produced by the Hoover Dam in an entire year. Where did it get it from? The answer is found in the stranger-than-you-might-have-thought subject of the rest of this chapter: *Gravity*.

Elementary, my dear Aristotle

On one fairly undeniable level, gravity is really a very straight-forward phenomenon indeed: when we drop an object, it falls. Children know and understand that this happens even before they can speak. It surprises no one. In that sense, gravity is one of the most basic facts of life.

In fact, our minds and bodies seem to be pre-wired to deal very effectively with gravity's effects. We hardly have to think about them at all, yet we can throw and catch, we can jump over things and land again without too much problem – some talented individuals can even juggle. The oh-so-familiar pull of gravity, in this sense, is *simple*.

However, as the mighty Greek thinker Aristotle (384–322 BC) pointed out in the first sentence of his magnificent *Metaphysics*, 'oh-so-familiar' isn't always enough: 'All human beings by nature desire to *know*.'

Sure, Aristotle says, we might be able to live quite functionally with gravity as an ever-present partner; but something in our human nature doesn't *let* us. Instead, we are driven to tackle even the seemingly 'obvious' by asking more questions. Aristotle himself is one of the greatest examples of this truth and it is difficult to think of a topic that he did not seek to address somewhere in his writings. So, naturally, he took on the physics of motion – and, with it, of gravity. What did he manage to come up with?

Building on the earlier theories of Empedocles (494–444 BC), Aristotle linked motion to the four elements: earth, air, fire and water (see Plate 1). All materials in our fallen and spoiled world were thought to be made of these four elements in various combinations; the heavens, on the other hand, were composed of the flawless *quintessence* or fifth element.

Beginning with this as his foundation, Aristotle aimed to show that gravity was simply a result of the elements associated with a material: the more earth and water in its essence, the more it was pulled towards the ground, its natural home. Conversely, those with air-and-or-fire as their dominant essences were drawn upwards.

This model, he claimed, described the *natural* motion of objects – as opposed to their *violent* motions. The latter, according to his theory, were movements caused by forces deliberately exerted by living things. The combination of natural and violent motion would therefore describe the overall path of the body.

This was not a time of carefully measured mathematical and experimental science, so Aristotle's claims would not be held up to anything more than what we might think of as a fairly basic observational test. If we apply his theory to our Tunguska explosion, for example, it broadly fits.

In essence, the comet fragment is earth and water, so its natural motion is towards the surface of our world. What's more, it is big and it is heavy, so it should head in that direction very quickly – which is precisely what it did. Aristotle, therefore, has provided a working model of gravity.

It's all Greek to me

Aristotle's ideas about the natural behaviour of bodies was a small part of his all-encompassing world-system – a system that was considered sufficient until surprisingly late in history. In fact, when Galileo was forming his own theories about motion nearly two millennia later (and threatening to undermine the entire status quo in the process), the main opposition came not from Catholic theologians, but from professors who made a comfortable living teaching Aristotle's physics.

Of course, with the benefit of both hindsight and the so-called scientific revolution of the 1600s, it is pretty easy to pull Aristotle apart. So easy, it would seem, that it is now the done thing in books that track the development of science. After all, the ancient philosopher was wrong about the five elements; he was wrong about celestial spheres; he was wrong about natural and violent motions; he was wrong to claim that heavier objects fell faster than lighter ones. This can begin to make him sound rather dim. Hawking, in *The Grand Design*, gives us an example of this kind of criticism:

He suppressed facts he found unappealing and focused his efforts on the reasons things happen, with relatively little energy invested in detailing exactly what was happening. Aristotle did adjust his conclusions when their blatant disagreement with observation could not be ignored . . . to explain the fact that objects clearly pick up speed as they fall, he invented a new principle – that bodies proceed more jubilantly, and hence accelerate, when they come closer to their natural place of rest, a principle that today seems a more apt description of certain people than of inanimate objects.[3]

The danger, though, is that we can go too far when highlighting the errors of our forebears.

Hawking probably gets his ideas here (albeit unwittingly) from B. F. Skinner (1904–1990), the celebrated Harvard psychologist and behaviourist, who has Aristotle allowing inanimate objects to think their own thoughts about how to move. This might not be a fair representation, though. Christopher Shields, a professor of Philosophy at Notre Dame, fights back on Aristotle's behalf with some strong words of his own:

> To anyone who has actually read Aristotle, it is unsurprising that this [Skinner's] ascription comes without an accompanying textual citation. For Aristotle, as Skinner would portray him, rocks are conscious beings having end states which they so delight in procuring that they accelerate themselves in exaltation as they grow ever closer to attaining them. There is no excuse for this sort of intellectual slovenliness.[4]

We should not be too hard on either Skinner or Hawking, though – for figuring out exactly what Aristotle *did* think is a tricky thing to do. His writing is not always clear, the precise definitions of some of his key terms ('weight', for example) are still argued about and he is not here to explain himself. While some suggest that Aristotle was really just anthropomorphizing objects rather than doing proper scientific thinking, Shields disagrees. He is insistent that, in correctly understood Aristotelian physics, stones and clouds do *not* move under their own will and they *do* obey fixed laws – but he also admits that the passages that prove this are complex and difficult to translate.

And yet, despite its limitations, difficulties and flaws, Aristotle's natural-motion explanation of gravity was good enough to satisfy the majority of intellectuals working on the problem until nearly the eighteenth century. Perhaps we should not be surprised, then, that its eventual and comprehensive unseating would have to wait for what may well have been the greatest mind of them all: that of the extraordinary Sir Isaac Newton.

This most beautiful system

Isaac Newton (1642–1727) is a far more fascinating character than many people realize. It is easy to form a picture of him as a scientific mastermind who calmly calculated his way to fame, discovering the mathematics of the universe along the way in an unemotional and highly rational manner – as a brilliantly clever, self-assured and enlightened English gentleman. Indeed, his famous words in a letter penned to another remarkable intellect, Robert Hooke – 'if I have seen further, it is by standing on the shoulders of giants'[5] – merely serve to reinforce the view that Newton was the perfect humble genius.

The reality – as we have already seen with Aristotle – is more complicated. At times, Newton was quite the bully. At others, he appears a fragile and broken victim. A towering figure who held a variety of influential posts during his career, he often behaved in ways that betray an intrinsic and destructive insecurity. Thus it was that Hawking, in his own introduction to Newton's physics, says:

Hooke challenged Newton to offer further proof of his eccentric optical theories. Newton's way of responding was one he did not outgrow as he matured. He withdrew, set out to humiliate Hooke at every opportunity, and refused to publish his book, *Opticks*, until after Hooke's death.[6]

As it happens, Hooke and Newton became bitter enemies and more than a few have suggested that Newton's 'shoulders of giants' comment to the former was actually sarcastic – Hooke was very short. The argument referred to in the quote above (for there were plenty of other disputes between the two) was over the nature of light itself. Was it composed of *waves* (as Hooke thought) or of *particles* (Newton)? While this discussion will prove immensely important to us in later chapters, for now we must stick with Newton's *gravity*.

It would probably be fair to say that the previous two centuries had paved much of the way for someone to come along and do what Newton did. This fact should not detract from his achievement, which was remarkable, but many of the key ingredients Newton used to formulate his theories had appeared only relatively recently. Special thanks for these must go to Nicolaus Copernicus (1473–1543), Johannes Kepler (1571–1630) and Galileo Galilei (1564–1642).

Building on an increased use of mathematics in the physical sciences, these three men each laid down hugely important markers for Newton. First, Copernicus reimagined our solar system by asking what it would look like if we stood on the Sun; he realized that doing so gave us all the known planets orbiting in the right order.

Second, Kepler carefully studied these orbits and noticed they were not the perfect, flawless, Aristotelian circles everyone believed in, but instead *ellipses* or slightly squashed circles.

Third and finally, Galileo put together new laws of movement that tore the heart out of Aristotle's natural and violent motions, mathematizing the paths of bodies in the presence of forces and providing a whole new structure for physicists to work with.

Alongside all this, Hooke and others had begun to discuss gravity itself – where was it coming from and how did it work? Hooke suspected that a certain mathematical relationship played a part, known as an *inverse-square law*. Put simply, this means that Hooke's gravity (which he thought emanated from the Sun) would not be twice as weak at twice the distance, but *four* times weaker; at *three* times the distance, it would be *nine* times weaker; and so on.

Hooke apparently believed this to be true because of Kepler's work – and he told the astronomer Edmund Halley (1656–1742) so in 1684. In what would prove to be a fateful decision, he also refused to go into any more detail because he had not published yet, so the frustrated Halley hot-footed it to Newton instead. Hawking takes up the story:

> [Halley asked] 'What would be the form of a planet's orbit if it were drawn towards the sun by a force that varied inversely as the square of the distance?' Newton's response was staggering. 'It would be an ellipse,' he answered immediately.[7]

All the pieces were now turned face up, ready for Newton to put into place. Encouraged (and financed) by the excited Halley, Newton wrote his masterpiece, *Principia*, in less than two years. In it is all the physics and maths needed to describe the motion of any object moving under gravity – be it an apple or a planet.

No quintessence was needed: Newton's laws were universal and applied equally to both the Earth and the heavens. His forces could act at a distance – even across space itself. He had cracked it. The equations were telling the story and Newton saw God at the heart of it. In his own words, at the end of the game-changing *Principia*: 'This most beautiful system of the sun, planets, and comets, could only proceed from the counsel and dominion of an intelligent and powerful Being.'[8]

Yet, although his formulae could describe the system – and all others like it – pretty comprehensively, what Newton really lacked was a *mechanism*. He looked for it all of his life and never found one. In other words, his maths could tell us how hard the Tunguska comet fragment would be pulled towards the Earth, the path it would take, the speed it would reach and how much damage it would do – but it could never tell us *why* it happened in the first place.

Personally persuaded that there should be particles involved in 'carrying' his forces from one object to another, Newton tried to find a way in which a particle-stream could head out through space and give rise to his equations. It was to no avail. In a letter to a friend, he admitted – once again with more than a hint of insecurity – that: 'Gravity must be caused by an agent acting constantly according to certain laws, but whether this agent be material or immaterial, I have left to the consideration of my readers.'[9] And, for the next two hundred years, none of them would work it out either.

Force fields and the art of science

We shall pick up the trail, then, well into the nineteenth century – with gravity simultaneously solved and not solved. Newton's Laws of Motion have proved robust and powerfully predictive, giving folk the ability to describe the motion down to a tee. All they had to do was plug numbers into an equation; the answer would obediently pop out. Isn't this the whole goal of science? Couldn't everyone just go home now?

The problem is the one Aristotle raised all that time ago – by nature, we desire to *know*. Yes, Newton gave us a mathematical solution, and very useful it was too. The Tunguska event was, on that level, simply a case of a large lump listening to laws. But *why* does gravity work? What *is* it?

Ultimately, we are asking if an abstract mathematical equation devoid of any mechanism is truly *enough*. Physicist and Newton scholar Richard Keesing thinks not. He, like so many others, does not want to stop with pragmatic calculation. Instead, he challenges us to think more deeply: 'What do you actually mean when you say that you *understand* something? You *think* you understand it; then you look more closely and it just evaporates!'[10]

This is precisely the mentality that drives science (and so many other disciplines) on – and there are very few periods in history when science grew faster than in the nineteenth century. Central to this was a flurry of experiments and theories surrounding what would eventually become the most world-changing of all scientific discoveries: *electromagnetism*.

A series of inventions and discoveries over the period from 1780 to 1820 meant that experimentalists were now able to play around to their hearts' content (provided they had wealthy backers) with electric circuits – artefacts that had been wholly unknown to earlier generations. One of these tinker-thinkers was the quirky and magnificent Michael Faraday (1791–1867).

What makes Faraday particularly unusual is that he was *not* a mathematician; in fact, he didn't like maths much at all. He had not received a formal education and spent his youth working manually as a bookbinder – which helps explain his highly individual hands-on approach. For Faraday, true science was physical, not abstract, and he wanted to work at a lab bench, not in a library.

As he experimented with magnets and electricity, he looked for relationships between the two. Sure enough, wires carrying electric current could be made to attract or repel each other, even at a distance. They could also affect the movement and behaviour of magnets – and vice versa.

Over time, Faraday catalogued hundreds of results and puzzled over how they could best be represented. Had it been Newton's project, a plethora of equations and calculations would have been the likely output, but Faraday thought differently. For him, the clear way to go was with *diagrams* – and the pictures he drew would change science for ever.

Faraday began to imagine contour-like lines emerging, almost ethereally, from his magnets and wires – lines that could then exert forces on other magnets and wires. Faraday's grand picture-world is a busy one in which these otherwise-invisible lines of force flow and move, fill up 'empty' space and push and pull other objects as they do so. He called his collections of lines – which, in combination, gave instructions on how everything else around them should behave – electromagnetic *fields*. There was hardly an equation in sight; the pictures were all the explanation he needed.

Field theory, as it became known, was no less successful for it – Faraday built the first working electric motors and even those we use today operate on the same principles as he drew in his sketches. Faraday's no-maths mentality is such a fine match to electromagnetic phenomenon that there is almost a ring of destiny to his story. The mathematicians would get their turn soon enough, yes – but Faraday equipped them with ideas that they would never have dreamt of themselves.

What has all this talk of magnets, however, got to do with *gravity* or Siberian comet fragments? Did Faraday take these on too? Perhaps surprisingly, the answer is no. Despite his firm conviction that all areas of science were deeply and profoundly interconnected, Faraday only considered gravity to be a small strand in this web and it simply didn't interest him much. So why mention him in this chapter at all?

Amazingly, it later turned out that Faraday's drawings of electromagnetic fields could work for gravity too. Instead of his lines of force emerging and ending on magnetic poles, they could be redrawn as beginning and ending on objects with *mass* – such as comets or tennis balls or neutron stars. In all other respects, his theory is kept essentially the same, yet its predictions of motion and forces now match Newton's mathematical work perfectly. Without either knowing it or intending it, Faraday's diagrams reimagined the rigorous equations in *Principia*.

In the light of this remarkable field-and-equation hybrid, the Tunguska event has a new explanation. Invisible gravitational field lines emerging from the passing comet met with those reaching out from the Earth into space and the two masses then danced to the resulting picture-music, following Faraday's carefully laid out score. Underneath his melody sits the rhythmic percussion of Newton's Laws – for the two theories are complementary ways of viewing the same interaction. They work brilliantly together, giving near-exact answers, and are now taught in tandem on every physics course worldwide.

Do these fields, then, finally give us the *mechanism* Newton craved? Do they provide the *understanding* that Keesing deems so elusive? Is Newton's gravity *caused* by Faraday's force fields?

One man – an unfashionable outsider unfancied by school and university – was prepared to suggest otherwise. His name was Einstein.

Light and momentary troubles

Albert Einstein (1879–1955) did not amount to much at school and decided to train as a teacher. Once again, he fell short and ended up in an administrative role in the patent office of Bern, Switzerland – a job he was awarded at least partially due to a friend's connections. There are simply no signs in his early life that he was about to become the most famous scientist in the world. As Hawking puts it:

> Genius isn't always immediately recognized. Although Albert Einstein would go on to become the greatest theoretical physicist who ever lived, when he was in grade school in Germany his headmaster told his father, 'He'll never make a success of anything.'[11]

Working at a time when many in the trade thought physics had nearly finished its task of describing our universe and would be complete in a matter of just a few years, Einstein undermined this belief – he penned several papers that turned the discipline upside-down and inside-out. We will discuss more of his work in future chapters, but it is his novel approach to our current topic of gravity that will occupy us for the rest of this one.

Like Faraday, who eschewed maths whenever possible, Einstein was an atypical thinker. As he later stated: 'I very rarely think in words. A thought comes and I may try to express it in words afterwards.'[12]

At the heart of Einstein's treatment of gravity lay a radical new assumption about how the universe itself works – one that no one further than a generation back could possibly have even imagined. To understand this, we need to consider what we mean when we talk about 'space' and 'time'.

Newton states near the beginning of *Principia* that, as he discusses the motion of objects through space and time, he will consider these two to form a rock-solid, unchanging backdrop on which events can play out. He says that space 'remains always similar and immovable'. He says that time 'of itself, and from its own nature, flows equably'. He says of both that they are like this 'without relation to anything external'.

These statements are integral to the rest of Newton's arguments. In fact, the entirety of Newton's gravitational system begins with concepts of *absolute space* and *absolute time* – in other words, locations and times are a guaranteed point of agreement for any observer looking at any problem.

This idea of a static background seems so obviously true that it might even seem strange for Newton to specify it – indeed, questioning it would be absurd. Einstein, though, was never bothered by what was considered absurd. In many senses, his fame comes from proving – on a number of separate occasions, no less – the absurd to be physically *true*. In this particular case, he decided that space and time might not be quite so fixed.

Intriguingly, we can track Einstein's space–time revolution all the way back to Faraday. The mathematical genius James Clerk Maxwell (1831–1879) had got involved – he took Faraday's field theory, combined it with other discoveries in electromagnetism and miraculously managed to reduce the sum total of everyone's findings down to just four beautiful formulae.

Faraday initially expressed reservations about his diagrams being mathematized, but eventually saw the value of it even if he personally did not like it. The response from the rest of the industry was far more positive; Maxwell's new maths ushered in huge theoretical and technological breakthroughs leading, ultimately, to the digital age. The cosmologist and science popularizer Carl Sagan sums it up nicely: 'Maxwell's equations have had a greater impact on human history than any ten presidents.'[13]

When Maxwell got to the point of sitting down and actually solving these equations, he realized that they would give rise to *waves*. These, remarkably, would manifest as ripples spreading out through Faraday's invisible fields. More remarkable still, he was able to calculate the *speed* these waves would travel at – and found it to be the speed of *light*. Amazingly, Maxwell had managed to discover the nature of light itself, a question that had been puzzled scientists and philosophers for millennia. His answer: light is an electromagnetic wave.

Einstein had been mulling all this over – and something was niggling at him. What would happen, he wondered, if he was able to ride next to one of these waves, matching its speed? What would it *look* like? Would it appear 'frozen' somehow? If so, something was wrong, because it didn't seem that any known science allowed such a frozen light beam to exist. On top of this, Einstein's gut told him there was no such thing.

Now, though, he had a problem. The only apparent alternative was the wave still getting away from him at light speed. But how could that be? If *he* was travelling at the speed of light, the beam would have to be going *twice as fast*.

Here, of course, things begin to unravel – for the natural next question is to ask what someone else watching all this from the side might see. Well, as far as they are concerned, the beam Einstein is trying (and failing) to catch is just a regular light wave: so it would travel at the speed of light, not double it. Oh dear. Exactly how fast is this beam travelling? It seems to be moving twice as fast as itself. Disaster.

A space-, time- and mind-bending solution

How could this nonsense be resolved – if at all? Einstein decided to do the unthinkable. He tore up (more accurately, he screwed up) Newton's (and everyone else's) ideas of space and time. No longer would they be the unchanging, fixed and independent background for objects to play out their roles upon. Instead, he would bring space and time on to the stage themselves – as *characters*.

Space and time, Einstein said, *could* change. Two people could watch the same series of events and disagree on where they happened or when they happened or even in what order they happened. In Einstein's world, the only constant was the speed of light, and all observers would agree on that.

Thankfully, he was also able to ensure that they would agree on the *laws* of physics – everyone would see a bouncing ball behave in the same general way, for instance. In his scheme, the arguments would come about the size or speed or location or duration or even colour of the ball and its bounces. For these, it really did matter where one was and what one was doing – all was *relative*. For this reason, Einstein's work is now referred to as *relativity*. Space and time weren't fixed – they were in the eye of the beholder.

Following the dropping of this bombshell, Hermann Minkowski (1864–1909) showed that Einstein's maths could be interpreted as a weaving of the newly flexible space and time together into a single sheet. Since there are three dimensions of space (height, depth, width), when these were combined with time, a four-dimensional surface emerged that we now call *space–time*. Space–time is impossible to picture in four dimensions, so all diagrams of it are drawn as a two-dimensional grid; but four-dimensional maths is not impossible, so calculations can be performed relatively easily.

But what of *gravity*? Einstein extended his model to include it in a new fantastic guise. In what became known as 'general relativity', he was able to show that wherever *mass* was present, it would cause the space–time grid to change shape. The greater the mass, the more it would bend space–time.[14] This simple assertion, it turns out, gives rise to gravity – but how so?

Imagine drawing a straight line on to squared paper and then gently folding the two ends of the sheet towards each other. In comparison to the grid, the line remains straight – but because the *paper itself* is now curved, the line is curved too. This, essentially, is how Einstein's version of gravity works.

Take the moon as an example: the reason it orbits the Earth is that the mass of the Earth has caused the space–time around it to curve inwards. In Newton's view, this couldn't happen, because space and time were absolute and independent – but, for Einstein, the Earth's mass directly affects the shape of space–time.

The moon, then, as it moves through space–time, follows this Earth-caused curvature – which explains its elliptical motion. Bizarrely, the moon itself is actually *moving in a straight line* – but it is a straight line through folded space–time. Which is, like the line on the folded paper, ultimately still a curve.

General relativity, therefore, provides a completely new way of thinking about gravity. In it, there are no longer gravitational forces or gravitational fields. Instead, there is only mass and space–time. In their odd relationship, mass tells space–time what shape to curl up into and space–time tells the mass what path to move along. Nothing is 'pushing' or 'pulling' anything else. There are no invisible lines of force. Einstein changed the rules – Newton's fixed background is gone. Instead, space–time is malleable.

It is this bizarre, stretchy, rubbery space–time that determines the movement of the heavenly bodies. It dictates the fall of an apple. It guides 50-metre-wide comet fragments travelling at thousands of miles an hour into the frozen Siberian wilderness. It even bends *light* and can cause changes in the passage of time.

What's more, general relativity has been proved true by experiments on time, distance, speed and space, over and over again. It offers minor corrections to Newton's previous predictions that turn out to be even more accurate than those of the great man; it makes astonishing new predictions beyond his, which have been brilliantly and dramatically verified; it solves complex gravitational problems that were impossible to pin down precisely in Newton's theory.

Einstein, then, had found more than just a mechanism for a centuries-old model everyone already knew was correct. He had found a new model, a stranger model, a *truer* model. As Keesing laments Newton's inevitable failure to calculate the exact orbit of the moon – a problem we now know was insoluble within his framework and that he died still trying to understand – he adds, with more than just a hint of regret: 'If only he had realised that gravity was the curvature of space–time!'

Hawking, relativity and those Big Questions

We have devoted a whole chapter to a single subject – gravity – and there is good reason for that: no book about the contribution Hawking has made to physics could avoid it. As a young man, he first made his name working on Einstein's general relativity – it underpins his PhD thesis and goes on to form the foundation on which he built all his most famous findings. Hawking is a relativist, through and through.

Because of this, being able to picture the warping of space–time is something that will matter on more than one occasion as we go on to analyse his theories and proposals. Black holes, Big Bangs, beginnings and other favourite topics depend on it. A whole new realm of possibilities is opened up by a bendable, four-dimensional universe; exotic ideas are suddenly freed to leap – with perfect legitimacy – from the pages of comics into mainstream scientific papers.

We promised the discussion of Big Questions in this book, at least partially because Hawking is constantly interacting with them. In this chapter on gravity, we have looked at the work of Aristotle, Copernicus, Kepler, Galileo, Newton, Hooke, Faraday, Maxwell and Einstein. What is fascinating about this list is that they – like Hawking – also felt compelled to write about God.

For Aristotle and Einstein, God is not quite so personal, but some sort of vague entity, or even vaguer 'oneness' that binds the universe together. For all the others, He is the Creator of the universe that they lived in and loved to investigate. Each, in fact, studied the Bible and found it to be a most helpful guide in their work. Each credited the Christian God with giving them motivation to do science and blessing them with insight along the way.

Each was also driven by their work to think about the Big Questions that Hawking himself raises. Towards the end of his life, Faraday was asked by a friend if he had any further speculations of note. His answer was to emphatically declare his faith in his Saviour: 'Speculations? I have none. I am resting on certainties. I know Whom I have believed.'[15]

In our next chapter, we shall see that general relativity was not the only mind-boggling paradigm shift in physics that Einstein helped to bring about. If anything, this second seismic change was bigger than the first, and stranger too. Just as with relativity, however, we shall need to review it – for it is the twin pillar on which Hawking bases his own view of life, the universe and everything.

As we shall see, the 'Grand Design' that he throws his considerable scientific weight behind sits atop not one, but two equally important columns: general relativity is the first – the other is quantum mechanics.

3

Canoe man • A case for the defence • The beginning of the end •
Hooke versus Newton revisited • But together they do • Young
strikes again • Weirdness, weirdness, everywhere • A matter of
interpretation • Heisenberg, histories and Hawking • Closing
argument

Canoe man

Paddling his canoe into the North Sea in 2002, John Darwin was
undeniably alive. Six years later, as he sat in the back of a prison van,
the same applied. It is his status *in between* these two events that
is the more unusual (and less obvious) one. During that interven-
ing period, as his struggling family would tearfully recount, he was
really not in a good way at all – he was *dead*.

Although the wreckage of his canoe washed up the day after
his death, Darwin's body was never recovered. His adult sons were
heartbroken at the loss of their father, but took a modicum of comfort from knowing that their mother, Anne, had not quite lost
everything. She received thousands of pounds of life insurance pay-
outs, and the policy paid off her mortgage too. Even the darkest of
clouds, it would seem, could still have a silvery lining.

Darwin was not officially declared dead until 2003. For a dead
man, though, he did tend to get about a bit. He was spotted by one
of his tenants later that year, then by a fisherman in Cornwall, then
by an internet 'friend' in Kansas, then by a boat dealer in southern
Spain. Clearly, none of these locations was exotic enough for his remarkably adventurous corpse, so it/he decided to fly to Panama in
the summer of 2006, with his bereaved wife coming along for the
ride. Once there, Anne and her dead husband looked at properties
for sale and even had their photograph taken with an estate agent.

Unsurprisingly, in an increasingly connected world, the Darwins' time and luck would run out and they began to realize the end was probably nigh. John decided to make a return from the dead and claim amnesia. Anne (who had recently moved to Panama) played along, expressing joy and wonder at the sudden reappearance of her husband in the UK. The police had already become suspicious of Anne at the time of her move, so the couple were now under double scrutiny. Was their incredible story true? Would Anne and John get away with it after all?

At this point, an interested member of the public stepped in. Gripped by the now international story of the resurrected 'canoe man', an anonymous woman decided to conduct her own investigation. Hilariously, she typed 'John, Anne and Panama' into Google and found the photo of the Darwins with the estate agent – complete with date.

The game was up. When Anne was shown the photo by the police, she surrendered. They had faked it all for the money, she said, and had even hidden the truth from their sons. John and Anne were packed off to jail and ordered to repay the more than half a million pounds they had fraudulently acquired. So much for silver linings.

A case for the defence

The Darwins were convicted of fraud. The reason for this was simple: they had claimed that John was dead, and he wasn't. For a while after his demise he had been living next door to Anne, and then had even moved back in with her. The eventual decision to try the amnesia route came because the paperwork had become too complex and they felt like it was all going to catch up with them. Once they admitted defeat, a conviction was rather straightforward. There was simply no case for the defence: dead canoeists don't buy yachts in Spain.

Underlying the court's final verdict, though, was an unwritten assumption: a person cannot be both dead *and* alive. A body decomposing in the North Sea cannot *also* be holidaying in Panama. One or the other, or even neither, is OK – but not *both*.

This particular assumption, however, was not debated in any of the sessions; John and Anne's lawyers did not challenge it or indeed even mention it. It is, after all, the commonest of common sense: bloated cadavers floating off Hartlepool can't visit associates in Kansas airports. Had the Darwins' counsel tried to claim otherwise, they would have been found in contempt of court.

Yet, in the weird world of Hawking's physics, this same logic does not always apply. For, in a run of successive scientific results around a century ago, this basic and so obviously true assumption – that objects can't be one thing and simultaneously the opposite or in two different places at once – was shaken to its very core. Eventually, to the shock and dismay of both the theoretical and experimental communities, it had to be abandoned entirely. Common sense, it would seem, was not quite as sensible as we thought.

Physics was transformed for ever. The old world, with its neat, tidy and well-behaved equations was gone. In its place was an evolving body of uncertainty and highly controversial formulae that no one truly understood. This was a new world unlike anything ever encountered before.

So great was the change that all physics from now on would be judged by whether it would fit into this radical regime or not, and this was even reflected in the language used. Henceforth, a theory was either *classical* – reasonable, regular and not subject to the up-setting, assumption-defying madness of the recent findings – or it was *quantum*.

The beginning of the end

As the nineteenth century rolled over into the twentieth, physicists were enjoying great success. They had conquered electromagnetism, made huge strides in thermodynamics, discovered X-rays and the electron, and many felt rather triumphant. The soon-to-become 'classical' physics was (mostly) going along very well indeed. Then Max Planck (1858–1947) came along and mucked it all up.

Planck accidentally ushered in the quantum age by tackling a well-known problem in a slightly unscientific way. The *black body radiation* puzzle had sat there as part curiosity, part frustration for a little while already before he decided to take it on. Essentially, the physics of the time predicted that a hot object would radiate an infinite amount of energy – a baked potato would destroy the known universe. As this was clearly wrong, the theory needed to be fixed. But no one knew *how*.

Rather than dealing with the actual science involved, Planck simply rewrote the equations again and again until they matched reality, paying no attention whatsoever to whether his maths made any sense or obeyed any established principles. As such, he made no claim that his new formulae had any real meaning or were even physically correct; they simply gave the right set of numbers and predicted how *real* hot objects behaved.

He didn't realize the significance of what he had done. A decade later, Einstein (and then others) would take his work and run with it, eventually giving rise to full-blown quantum mechanics. Planck's pragmatic-but-unreal maths turned out to be a description of what had, until now, been the completely hidden foundation of our subatomic physical existence. The cat was now out of the bag – and a very strange cat it would prove to be.

Hooke versus Newton revisited

A century after Hooke and Newton had tangled over the nature of light, the issue was put to bed by Thomas Young (1773–1829). He performed a very clever experiment that proved, definitively, light is a *wave* – Hooke was right after all. This is not Young's only achievement of note: he also developed a measure of material strength still used by engineers today, he introduced the word 'energy' into modern physics and he was the first person to decipher Egyptian hieroglyphics. In short, he was a bright chap.

Young's set-up was rather simple (see Plate 2). He shone light towards two narrow slits and observed the resulting pattern on a screen. Had light been composed of particles (as Newton believed), then the expected outcome was two clearly defined bright patches, each of which lined up directly with a slit. That was not what Young ended up seeing.

Instead, the screen displayed a series of alternating bright and dark fringes. This is precisely what would be expected if light travels as *waves*. As each wavefront emerged from its slit, it would spread out in a semicircle, like ripples across the surface of a pond. The two sets of ripples would then overlap in the space between the slits and the screen.

When waves like this overlap, they can add up (if a peak meets a peak) or they can cancel out (if a peak meets a trough) and they do so in a neat, geometric fashion. This gave rise to the bright fringes (waves adding up) and the dark fringes (waves cancelling out) in different locations on the screen. Young had done it. The answer was finally beyond doubt: light is not made of particles. Light is a type of wave.

Young's result held good for a hundred years until, like Newton, he fell victim to the all-conquering Einstein. Once again, a flash of wonderful insight from the German tore up the rulebook. What if Planck's crazy make-it-up-until-it-works maths was more than just a fudge, Einstein asked? What if he had actually uncovered the *truth*?

Einstein played around with Planck's model and found that he could use it to explain another discrepancy in physics, known as the *photoelectric effect*. This was an easy-to-perform experiment in which light of different colours was shone on to charged metal plates to note its effect. Everyone could get the results, but no one could say why they were happening. When Einstein treated Planck's work as real, however, he could show why the effect occurred – and it won him the Nobel Prize.

The fact that Planck's strange maths had now shown up in two different experiments made people sit up and take notice. With Einstein's claims about the photoelectric effect, though, came a new postulate: a light beam might not be a wave after all. Instead, it could be thought of as a stream of energy packets or *quanta*. These packets were, essentially, *particles*; they were soon given the name *photons*. Newton, it would seem, had been right all along.

By 1905, then, physics found itself in a very odd position indeed. Young had proved light was a wave; Einstein had proved light was a particle. Both results were theoretically sound and both were supported by well-known experiments. How could a deadlock like this be broken? What was going on? The answer was crazier than anyone might have dreamt – and was worth quite a few more Nobel Prizes.

But together they do

Louis Victor Pierre Raymond de Broglie (1892–1987) sounds more like a figure from an Alexandre Dumas novel than a candidate for solving our wave–particle dilemma. A wealthy lifelong-bachelor, the Seventh Duke de Broglie began his academic life as a historian – a fact that makes what was to come even more surprising. In his PhD thesis of 1924, he dared to say something out loud that other quantum trendsetters – although they may well have been privately pondering it – were yet to voice.

Maybe the debate in hand is too small, he said. Maybe Einstein played it too safe by only assigning this weird particle-behaviour-of-a-wave to light. Maybe it was time to stop defining *any* type of natural phenomena – light, atoms, Panamanian estate agents – as *either* wave *or* particle. Instead, de Broglie insisted, there was a third way. They could be *both*.

If this sounds borderline insane, that's because it sort of is. Just a few years earlier, de Broglie would have been laughed out of the lab if he had dared to propose something so anti-common sense. The thing is, though, that post-1905 physics had developed its very own 'anything goes' mindset – the quantum results flying around were so inescapably mad that there was a new willingness to consider even the most outlandish of ideas.

In this case, de Broglie had simply chosen to extend Einstein's original Planck-based concept far beyond its original domain. Einstein had already argued that wave theory and particle theory could operate hand-in-hand when necessary – in *A Stubbornly Persistent Illusion* (a volume edited, incidentally, by Hawking) he is recorded as saying:

> But what is light really? Is it a wave or a shower of photons? . . .
> It seems as though we must use sometimes the one theory and
> sometimes the other, while at times we may use either . . .
> separately neither of them fully explains the phenomena of
> light, but together they do.[1]

If a 'wave' could sometimes act like a particle, de Broglie believed, a 'particle' could act like a wave. In his 1924 paper, de Broglie declared his hand: he thought that Einstein's principles also applied to electrons, as well as to anything else that had a mass and a velocity. As a result, he was saying that we should be able to find *electrons acting like waves* out there somewhere. Within three years, experimentalists found precisely that – and the noble got a Nobel.

Young strikes again

The electron is a funny beast. Discovered in 1897 by Joseph John Thomson (1856–1940), it was found to have very specific and fairly peculiar properties. It was smaller than an atom; it emerged from within atoms; it had a fixed mass and a fixed electric charge. Quite why it has these properties remains a mystery even today, but physicists still hope an ultimate theory might tell us – as Hawking says in *The Grand Design*, 'Many scientists believe there exists a single theory that explains [nature's] laws as well as nature's physical constants, such as the mass of the electron.'[2]

Electrons are everywhere – all atoms contain them, so computer punch-cards, Siberian comet fragments, wrecked canoes and even John Darwins are full of them. In that sense, they are really quite ordinary. In 1927, however, an experiment was carried out in which they proved to be anything but. It is a strong candidate for the oddest scientific result of all time – and it led, pretty much directly, to most of the other candidates too.

We have already discussed Thomas Young's brilliant analysis of light, in which he passed beams through two slits, showing that they spread out and overlapped just like waves. In 1925, New York physicists Clinton Davisson (1881–1958) and Lester Germer (1896–1971) managed to produce the same effect – but this time with a beam of slow-moving *electrons*.

Before we explain the nature and consequence of their discovery, however, it is worth relating how they actually arrived at it. It is a fascinating story in its own right – but it also provides an enlightening glimpse of what *real* science can be like. In particular, it demonstrates how the famed 'scientific method' is not often as clear-cut or clinical as outsiders are led to believe.

First, Davisson and Germer were not actually looking for electron waves; they had no initial interest in the work of de Broglie at all. Instead, they were working on something entirely different – analysing the structure of nickel. In the annals of science, however, it is not uncommon for major discoveries to come along uninvited and unexpected, so, strange as it may seem, this surprise is not all that surprising.

Second, an accident with their equipment damaged their metal sample; when they tried to fix it, they unwittingly changed its composition. If this had not happened, there would have been no quantum breakthrough. Again, though, this is not as rare an occurrence as one might have thought: apparent scientific misfortune frequently brings forth new scientific understanding and frustration gives way to elation.

Third, when they actually got their results, they had no idea what they meant – they only discovered the wave–particle link by chance. While on holiday in Oxford, Davisson made the impromptu decision to attend a lecture given by quantum pioneer Max Born (1882–1970) – and he was astonished to see that his own data was being used as potential proof of de Broglie's ideas. Amazed, excited and more than a little confused, he returned home and proceeded to teach himself quantum theory. He went on to formalize his findings in 1927.

Davisson had indeed – despite a lack of intention and a few accidents, and thanks to a jaw-dropping coincidence – proved the wave–particle duality of the electron. His bungled experiment had become, in essence, an updated version of Young's slit experiment – and the electrons involved showed wave behaviour just as de Broglie had predicted.

For these more than slightly fortuitous efforts, Davisson received the 1937 Nobel Prize. He had also, as we shall soon see, opened up a can-full of quantum worms.

Weirdness, weirdness, everywhere

What, exactly, does the wave–particle duality of the electron *mean*? One way to think about it is this: the pattern that Davisson and Germer picked up on their screen was basically the same that Young had picked up on his, so the electrons were acting like waves (see Plate 3). However, it was undeniable that there were then *individual* electrons hitting the detector – which is particle behaviour. The conclusion is a strange one indeed: the electron switches from particle (when released) to wave (when passing through the slits) to particle again (when striking the screen).

It gets odder. Cover up one slit, and things change: Young's familiar pattern disappears and the electron acts as a particle all the way through. Somehow, the two-slit set-up itself triggers the wavelike behaviour. Could this be because when multiple electrons travel through two slits, they can interact with each other, but can't do the same through just one? A simple way to test this is to release them *one at a time*, letting each hit the screen before the next is sent on its way. What happens then?

It gets odder again. Even when released separately with a decent amount of time between them, the double slit results in the wave pattern, but the single slit doesn't. This means that an electron appears to 'know' whether there are two slits or one, then acts accordingly – but how could that possibly be the case?

More than that, if these distinctive patterns are produced by overlapping waves emerging from the two gaps, we are forced to a seemingly nonsensical conclusion: a single electron somehow passes through *both* slits simultaneously, then spreads itself out and then – in some bizarre fashion – overlaps *itself*.

So crazy are these findings that Matt O'Dowd, astrophysicist and science popularizer, has dubbed this 'the quantum experiment that broke reality'.[3] Since the time of Davisson and de Broglie, though, it has been repeated on numerous occasions, with many variations, and each time the conclusion is the same. But how can electrons be in more than one place at the same time? How can they go through the left slit *and* the right slit? How can they interfere with themselves?

A matter of interpretation

While the new field of quantum mechanics (QM) was birthed by experiment – black body radiation, the photoelectric effect, the electron double slit – it is, at heart, a mathematical theory. There is a plethora of equations, many of which are fearsomely complex to even the brightest of minds, but the solutions they provide match up with experiment again and again. In fact, one of the first self-contained models within the quantum world, *quantum electrodynamics* or simply QED (which we will study later in this book) is the most accurate science ever devised.

Working with these equations, however, has thrown up all sorts of ideas that no one was really prepared for. For a start, they turn any scientific measurement into a *probability* rather than a certainty, and therefore allow – in theory – for *anything* to happen.

Take the following as a suitable example: if QM were to analyse the moon, it would genuinely include in its formulation the possibility that it is made of cheese; or will suddenly turn into cheese tomorrow; or will switch between cheese and rock every time Tottenham Hotspur wins the league. These probabilities are very small (especially the Tottenham-winning-the-league one) and are vastly outweighed by the prediction that we will just keep the same familiar moon for the time being – but they are, undeniably, *there*.

In fact, QM places extraordinary value on the *act* of measurement itself. Until a measurement is actually made, it suggests, the universe appears to contain *all* the possible results – however unlikely they are – simultaneously. Only when the reading is taken will the universe reveal a single definitive answer that will (usually) be the one with the highest quantum probability.

This has disconcerting repercussions. In our example above, for instance, the maths seems to say that the moon *is* made of cheese *and* it isn't, *and*, even, that it is currently hosting a welcome home party for John Darwin – all until we bother to *look* at it. Once we do that, the infinite sea of possibilities disappears and it just becomes the plain old moon again (probably).

As one might imagine, this caused quite some consternation in the physics community. How could these wildly different measurement outcomes all 'exist' at the same time? Why did they nearly all suddenly vanish to leave just one rather ordinary outcome whenever someone cared to pop in and see what was going on? Surely the moon is still rock when no one is looking at it – isn't it?

The real heart of the problem is this: the sort-of-wave-sort-of-particle-maths of quantum mechanics *works*; but we don't understand what it *means*. This is precisely why it generates so many quotes from top physicists – including its own discoverers – about how ridiculous it all is. Even the brilliant Richard Feynman (1918–1988), who was one of the key players in the game and who laid some of Hawking's most important foundations, was stymied. Here is his bald assessment: 'I think I can safely say that no one understands quantum mechanics.'[4]

As a result – and as un-sciencey as it might seem – there are now more than twenty different interpretations of QM doing the rounds,[5] and the professionals are no nearer to agreement now than they have ever been. Professor Sean Carroll, cosmologist (and genuinely excellent science writer, as it happens) describes the situation this way:

> Think about it – quantum mechanics has been around since the 1920s at least, in a fairly settled form . . . without it nothing makes sense. And yet – we don't understand it. Embarrassing. To all of us, as a field (not excepting myself).[6]

The first attempt to explain what QM means – the *Copenhagen interpretation* – arrived in the 1920s. It says that all the possible states of a system really do coexist as a combined mathematical entity called the *wave function* until someone measures it. The measurement makes the multi-value wave function 'collapse', leaving behind just one state from the previous mix, which is otherwise known as the actual result.

The quantum maths involved in the wave function tells us the *likelihood* of each possibility being the one that survives; but it does not, therefore, give one definitive prediction. Under the Copenhagen interpretation, the measurement actually *causes* the result – it would not exist without it. In other words, the observation of reality is needed if there is to *be* a reality. There is no rocky moon, it says, until we *look* at it.

Rival *realist interpretations* (more on realism later on) claim that the moon is there all the time, whether anyone is observing it or not – sometimes by picturing the wave function as a real, physical wave of some kind.

Yet another class, known as *many world interpretations*, suggest that our measurements give rise to *every* outcome, with each being realized in an entirely new and entirely separate universe. This would mean that when we look at the moon, the universe splits into millions of different realities – one for each quantum possibility. If this is true, then the moon really *is* made of cheese in some of those universes, and John Darwin really *is* dead – billions of times over.

Interpretations of quantum mechanics, however, are a completely different type of science – that's if they are science at all. Every interpretation uses the same equations; each gives identical results. As such, they are physically indistinguishable and no (current) experiment can judge which is right. This means that the differing interpretations are, as bizarre as it might seem, simply a matter of taste. It does not necessarily follow that we will *never* know what the maths is really telling us about our universe – as we shall find out in later chapters – but, for now at least, we are well and truly stumped.

Heisenberg, histories and Hawking

We have barely scratched the surface of QM in our discussion above, but we have managed to cover a few crucial points. First, it generally applies to the very small – electrons and photons, for instance.

Second, it works – brilliantly. Third, it is weird. Fourth, it blurs boundaries between waves and particles. Fifth, it is wide open to interpretation. Much of Hawking's contribution to physics is tied up with quantum theory, so we will need to bear each of these facts in mind when we go on to look at his ideas.

Before we can do so, though, there remain two more quantum ingredients to briefly mention. The first is disturbing and, worse, turns out to be unavoidable. The second is a breathtaking leap of mathematical genius.

Werner Heisenberg (1901–1976) was pretty much there from the start. Along with the aforementioned Born, he deduced the first proper formulation of quantum mechanical ideas in 1925. Two years later, he dropped a bombshell that still shakes us today – his (in)famous uncertainty principle.

Put simply, Heisenberg identified pairs of measurements that were linked in a very specific way – the more accurately you knew one, the more uncertain you were about the other. If we knew the exact *location* of a particle, for instance, we would be clueless about its *motion*. If we try to resolve this by finding new information about its motion, we are, in turn, made less sure of its location. We can't have our cake and eat it.

There was no way around this: no cheating was allowed. Einstein hated the uncertainty principle and fought day and night against it, but his ideas were shot down again and again. We now know that Heisenberg's maths describes a fundamental truth about our world. And, as we shall see, it plays a major role in Hawking's greatest works.

Another hugely important quantum ingredient for Hawking is a mathematical trick devised by the maverick freethinker Richard Feynman. Mulling over the double-slit experiment, Feynman realized that he could dodge the horrendously complex equations by doing something very strange indeed: pretending that the electron had *multiple histories*.

Feynman was interested in calculating the probability that an electron launched from a particular point (A) would eventually hit the screen at a second particular point (B). Surprisingly, he found that if he imagined the electron taking *every* conceivable route from his A to his B simultaneously, the final answer was much easier to determine than by using previous techniques.

The difference between his approach (see Plate 4) and the conventional one is subtle, but it is worth noting that Feynman's electron does *not* turn into a wave at the slits. Instead, it remains a particle – but one that makes the A to B trip in an infinite number of different ways, all at once.

Incredibly, he put very few limits on what these journeys could look like – he included all paths through each slit or neither, detours through Himalayan monasteries, and trips to the end of the universe and back. Feynman was even relaxed enough to let the electron break the light speed barrier along its way.

It is very hard to see how imagining an infinite amount of electron histories could possibly make things *simpler*, but the beauty of Feynman's method is that, in the final summing up, nearly all the countless routes cancel each other out. Just one number remains – the probability of finding the electron at point B – a value that is in agreement with those from the older and more complicated methods.

Feynman, then, had invented a brilliant shortcut – one that opened the way for physicists to look at new problems and have a decent chance of cracking them. His *sum over histories* idea was a genuine game-changer – and Hawking later utilized it to fantastic effect.

Closing argument

The quantum world permits all sorts of things that are utterly alien to us on the larger-than-a-few-atoms scale. Had John Darwin been the size of a nucleus, his lawyers could have argued that, in between his disappearance and reappearance, he had been *both* alive *and* dead – and that Anne was therefore not really committing fraud at all.

Alternatively, they could have blamed the police for making a measurement when they interviewed him, and that they therefore *caused* his wave function to collapse into a state of resurrection so his living status was *their* fault, not his. Finally, they could have pleaded that John really *was* dead in a near-infinite number of other universes which had branched off from ours – and it was rather harsh to make a judgement purely on matters in this one.

Even though these weird and wonderful effects don't appear to manifest in our everyday lives, they are unquestionably there at the level of our building blocks. This leads to some uncomfortable questions. Are there really multiple universes in which anything that can happen has happened? Do things exist when we aren't looking at them? Is our world inescapably uncertain? What does all this mean about *meaning*?

In our previous chapter, when we reviewed the long history of gravity, we found that the key thinkers involved were also drawn to the Big Questions. The same is true of QM. Planck, Born, Heisenberg and others simply could not resist considering what this new science was telling them about life, the universe and everything – God included. Planck, who inadvertently kicked the whole thing off, said: 'Both religion and science require a belief in God. For believers, God is in the beginning, and for physicists He is at the end of all considerations . . . To the former He is the foundation, to the latter, the crown.'[7]

Likewise, in a lengthy passage, Heisenberg – who was awarded the 1932 Nobel Prize 'for the creation of quantum mechanics' – feels compelled to track the progress of human understanding:

The first thing we could say was simply 'I believe in God' . . . the next step was doubt. There is no God. And yet [today] we may with full confidence place ourselves into the hands of the higher power who . . . determines our faith and therewith our world and our fate.[8]

Hawking, as we might expect, also links QM with God – both on the scientific and the philosophical fronts. Yet why was he working with quantum mechanics at all? His interest, as we know, is in the universe as a whole. Isn't that scale unimaginably *big*? Why should he care about the very very *small*?

The answer is, the universe, now so indescribably huge, might not always have been that way – but that is the topic of our next chapter.

4

Gingerich's quest • All that nonsense • The book nobody read • Holding the universe still • The 1920s' Copernicus #1 • The 1920s' Copernicus #2 • Redshifting opinions • The discarded image • A cooling-off period • The small matter of nucleosynthesis • Digging the cosmological heels in • Brainwaves and microwaves • Cosmology and salvation

Gingerich's quest

Harvard University is the best in the world – at least, it is according to the highly regarded Shanghai Ranking system.[1] In 2018, it edged Stanford and Cambridge into second and third places, respectively. This was no flash in the pan for the Boston-based body, by the way; Harvard was number 1 in 2017 and 2016 as well. In fact, it has topped the rankings every single year since they began.

Attaining a major academic position at Harvard, therefore, is quite some achievement; maintaining it for multiple decades is even more impressive. Professor Owen Gingerich has managed to do both. He has also chaired the university's History of Science Department, been senior astronomer at the Smithsonian and can even claim to have an asteroid – *2658 Gingerich* – named after him. Nearly ninety years old (at the time of writing), he remains one of the foremost authorities on astronomy anywhere in the world.

Yet, despite this mammoth list of achievements, Gingerich stands in awe of another. So much does he admire this fellow stargazer, in fact, that he has spent more than thirty years of his life seeking out and then promoting the other man's work.

His quest has taken him to various obscure libraries across North America, Europe, China, Japan and Australia; there was even a (successful) cloak and dagger attempt to obtain a crucial astronomical microfilm from the former Soviet Union during the height of the Cold War.

What astronomical result could possibly be worth chasing down with such bloody-minded determination? Who is this mystery man that even the all-conquering Gingerich is blown away by him? The answer lies in a scientific finding that no one was looking for; that no one wanted; and that no one suspected could ever be true – that the Sun is at the centre of our solar system.

All that nonsense

When the Polish churchman Nikolaus Copernicus (1473–1543) first suggested that the Earth might be circling the Sun, it is not even clear whether he himself believed it. For well over a thousand years, the motions of the heavens had been modelled according to the Earth-centred *Almagest* of Claudius Ptolemy (100–160) and all had been fine, thank you very much.

Copernicus, however, was frustrated. Ptolemy's system worked pretty well, but it was ungainly and inelegant – the Pole despised the mathematical fudges and awkward compromises that had been included to make sure it functioned properly.

Then, in a stroke of extraordinary genius, Copernicus realized that if he imagined he was standing on the Sun, he could solve many of these fiddly problems. From this new standpoint, he proved, the planets would now orbit in a more sensible sequence. Other (and vastly more technical) pieces of the puzzle fell into place too; this Sun-centred universe was less complicated – and it felt, to Copernicus at least, far purer.

The issue was, though, that his was a stupidly impossible suggestion. Heliocentrism (placing the Sun in the middle) required the Earth to be hurtling through space at breakneck speed. The atmosphere would have been blasted away and human beings would have been launched into limbo. Gingerich imagines being an astronomer at the time and hearing Copernicus's ideas: 'You would no doubt have told him to get lost and to take all that nonsense with him . . . think how much harder it would have been to walk west than walk east! Totally ridiculous!'[2]

Copernicus anticipated this response. For years, he did not publish his ideas in full – although he did circulate them in part in his *Commentariolus* (Little Commentary) some time around 1520. Eventually, however, a group of friends (that included a cardinal) managed to persuade him to print the whole show – Copernicus did so, but remained wary of just how crazy his ideas would seem to many. In his preface, addressed to the then Pope, he says:

> I can easily conceive, most Holy Father, that as soon as some people learn that in this book which I have written concerning the revolutions of the heavenly bodies, I ascribe certain motions to the Earth, they will cry out at once that I and my theory should be rejected.[3]

Later in the same document he seems to get braver – he defiantly stands his ground, declaring that he will remain unbowed by any criticism. He thinks his views are *both* scientific *and* Christian:

> If perchance there shall be idle talkers, who, though they are ignorant of all mathematical sciences, nevertheless assume the right to pass judgment on these things, and if they should dare to criticise and attack this theory of mine because of some passage of Scripture which they have falsely distorted for their own purpose, I care not at all; I will even despise their judgment as foolish.

The stage, then, was set. So what *was* the reaction when his dramatic and controversial new visualization of the universe hit the medieval bookshelves? Was he hissed off the stage as he feared? Was he celebrated as a magnificent astronomical genius? Was he burned at the stake by angry Catholics?

The answer is 'none of the above'.

The book nobody read

Tragically, Copernicus became ill during the printing of *The Revolutions of the Heavenly Bodies* and died on the same day that he was presented with the first edition. This has led to the convenient (to some) but wholly untrue notion that Copernicus was terrified of the Inquisition and only published when he already knew he was dying.

Instead, the general response to his work was rather muted – a strange and unexpected fact given (a) the inflammatory content of the material and (b) the worldwide fame of Copernicus nowadays. The flatness of its non-impact even prompted the prominent author Arthur Koestler (1905–1983) to dub it 'an all-time worst seller' because it was 'supremely unreadable'.[4]

Enter Gingerich, stage right. Inspired by a chance encounter with a centuries-old and heavily annotated copy of Copernicus's first edition, he decided to track down the rest of them – including one version held on a Russian microfilm. Again and again, he found that astronomers of the time *had* read (and written in) it. So why, if it was so very controversial, had there not been more of a reaction? Why had Koestler concluded that it was 'the book nobody read'? Gingerich has a theory:

Surely, Koestler must have reasoned, if scholars actually read the book, they must have promptly seen the light, and since widespread adoption of the new cosmology did not happen, it must have been a book that nobody read. It apparently never occurred to him that scholars read *De revolutionibus* as a recipe book and just did not believe that it applied to physical reality.[5]

Why would scholars think this? Well, many editions of Copernicus's text went out with a preface written by his overly cautious friend, Andreas Osiander (1497–1552), in which it was stressed that the new model, first and foremost, was a mathematical device. To think, however, that this was the only reason the scientific world didn't immediately adopt heliocentrism is grossly naive – for there are plenty more.

First, most thinkers at the time were quite content with Ptolemy's ancient Earth-centred model – they simply were not fussed about finding a new one. Second, Copernicus had very little evidence on his side, since there was neither sensation nor indication of the movement of the Earth. Third, his resulting mathematical tables could still be used to deduce astronomical positions without taking any notice whatsoever of where exactly he had placed the Sun.

In other words, his theory was not wanted, not needed and not believed. It had limited occasional use, yes, but no one thought for a second that it was *real*.

Holding the universe still

Leaving the late medieval cosmological shake-up behind for the moment, we shall jump forward now to the first decades of the twentieth century again – a time that we have already seen had been shocked by both general relativity and quantum mechanics.

Einstein, though, wasn't finished. Having formulated his theories of gravity, and proving they could solve difficult individual cases, Einstein became interested in what might happen if he applied them to the universe *as a whole*. Obviously, this is a horrendously complicated thing to do, as his model was all about mass bending space–time (see Chapter 2) – so he would need to include *all the individual masses everywhere in the universe.*

Since this was clearly impossible, Einstein simplified the situation by assuming that the universe was the same in every direction (isotropic) looking from anywhere (homogenous). His result, however, was that space–time always collapsed in on itself – gravity made it curl disastrously inwards no matter how the mass was arranged initially. Einstein's hitherto all-conquering model had been hiding a horribly fatal flaw – on the universal scale, it was hopelessly unstable. Relativity would eat itself.

To avoid this, the German visionary sneaked in an invented mathematical term called the *cosmological constant* – a wholly unexplained quantity that balanced the inward pull of gravity by bending the space outwards again. Thanks to this questionable fudge factor, Einstein's universe no longer caved in; it just sat there, uneventfully stable, for ever.

Cosmologists liked his solution, because they already believed in an eternally static world and had done for centuries. Einstein, however, was uneasy, because he knew that he had messed with his original and elegant formulation. The problem was that he could see no alternative – he *had* to force his space–time to stay still, just as everyone agreed it should. In a 1919 paper, he admits his reservations about tweaking his theory: 'this [forever stationary] view of the universe necessitated an extension of equations with the introduction of a new universal constant λ ... this is gravely detrimental to the formal beauty of the theory'.[6]

In the end, Einstein and his contemporaries simply could not envisage a cosmos that was anything other than still. Everlasting sedentary space was the scientific dogma of the day, having enjoyed millennia of assent – in precisely the same way that Ptolemy's fixed Earth had done. The 1920s, it would seem, needed their very own Copernicus to come along and challenge the stubborn and incorrect consensus – but would they get one?

As it happens, in fact, they did rather better than that. They got *two*.

The 1920s' Copernicus #1

Alexander Friedmann (1888–1925) was, much like Richard Feynman of *sum over histories* fame, a bit of a maverick – he relished going against the grain. In 1922 this Russian mathematician put together a completely new way of thinking about Einstein's unstable universe. What if, he asked, we didn't assume the cosmos was still? Why couldn't it be *moving*?

Wonderfully, Friedmann was able to demonstrate – in what might almost seem like common sense after the event – that the tendency of mass to pull space–time inwards didn't have to lead to a collapse. Instead, if he started with *an already-expanding universe*, all that the mass would do was *slow the expansion down*. In fact, he was able to find three different possible models of the cosmos, each one dependent on how much mass there was in total.

In the first, lower-mass version, the universe expands outwards for ever. This is because there is not enough mass to halt the growth in any significant way – and as space–time grows, the matter in it gets even more spread out and, consequently, the inward pull becomes weaker.

In the second, with a much larger concentration of mass, the expansion would slow, stop, and then gradually reverse – giving the dramatic universal crunch that Einstein had tried to dodge.

The third possibility was a very finely balanced middle option – expansion would continue eternally, but would get slower and slower and slower, nearly reaching a stop.

Friedmann was in for a kicking. His papers weren't circulated widely, they weren't read much, he had gone against the prevailing wisdom of the entire scientific community, and he had no physical evidence to support his work. Even when news of his moving universe did get anywhere, it was snuffed out almost instantly. Einstein, who might possibly have been able to rescue it, hated the idea – and wasn't afraid to say so in public. Simon Singh, in his highly entertaining book *Big Bang*, puts it this way:

His ideas had been published, but in his lifetime they were largely unread and completely ignored. Part of the problem was that Friedmann was simply too radical . . .

To make matters worse, Friedmann had been condemned by Einstein, the world's most prominent cosmologist.[7]

Friedmann died unexpectedly in 1925 without seeing the acceptance of his theories. Immediately, however, another rogue physicist took up his cause – this time a Catholic priest – and added a brilliant final flourish to the argument.

The 1920s' Copernicus #2

George Lemaître (1894–1966) came up with his idea of an expanding universe entirely independently of Friedmann. Unlike the Russian, though, he did not stop with just the maths. If the universe was indeed expanding outwards, he realized, then it must be getting *bigger*. This sounds blatantly obvious (a truism, even) but Friedmann had never considered it. This was because the Russian was a mathematician, interested only in the equations themselves. Lemaître, by contrast, was a *physicist* – he had a picture, in his head, of *reality*.

If the universe is getting bigger, he said, then *in the past it must have been smaller*. Go back further and it gets smaller still. Eventually, Lemaître reasoned, you would hit a *beginning*.

This insight was every bit as world-shattering as Copernicus moving the Earth. Everyone, scientists and all, *knew* that the universe was eternal; everyone *knew* it stood still. It was crazy enough to suggest it might be moving somehow; but to postulate a beginning was madness in the extreme.

Lemaître's controversial suggestion was, therefore, also utterly preposterous – we are reminded of Gingerich's 'nonsense'. Once again, Einstein was at the forefront of the catcallers, rebuffing Lemaître at a 1927 conference in Brussels. He told the priest, in no uncertain terms, that: 'Your calculations are correct, but your physics is abominable.'[8]

What Friedmann and Lemaître lacked – like Copernicus before them – was enough *evidence*. Could anyone offer any physical proof that their expanding cosmos might be *real*?

Redshifting opinions

One year before Einstein published his 1915 paper on general relativity, the American Astronomical Society met in Evanston, Illinois. There, Vesto Melvin Slipher (1875–1969) announced an extraordinary discovery – he had noticed that the light emitted from mysterious cloud-like objects called *nebulae* was nearly always *redder* than it was supposed to be.

Astronomers at the time had already figured out a way of measuring the speed that stars were travelling at – all by looking at the colour of their light. If a star was moving towards us, its light waves would be slightly 'squashed', which made them look bluer. The opposite was also the case: *redshifted* stars were moving away. What's more, the speed could be calculated from the colour change and, typically, was tens of kilometres per second.

Slipher, though, took this a step further – he applied the technique to nebulae rather than stars. What he found staggered him: these enigmatic blurs were moving much, much faster than any star did and they were *almost all moving away from us*. His presentation to the society was met with genuine astonishment; when he finished and sat down, the audience jumped to their feet and cheered.

Quite what it all meant, though, was not clear – not yet, anyway. Astronomers could now figure out how fast nebulae were moving, but they didn't, at this point, know how far away they were. Many considered the oddities to be part of our own galaxy; others argued that they were much more distant than that – and might even be galaxies themselves.

In the audience on that fateful 1914 day (perhaps he himself applauded Slipher) was a young Edwin Hubble (1889–1953), a man who would go on to become the greatest experimental astronomer of all time. In 1923, he proved that nebulae were indeed galaxies in their own right and, crucially, he could also figure out how far beyond our Milky Way they really were.

Putting all this information together in 1929, Hubble produced the evidence that would eventually swing the pendulum right the way over to Friedmann and Lemaître – because, in every direction he pointed his mighty telescope, these far-off galaxies were red-shifted. They were all moving away and the furthest were moving the fastest.

Run time backwards, it could be shown, and they would all meet at a single point. This was exactly what Lemaître needed – proof. Einstein, then, and pretty much everyone else as well, had been wrong all along. The universe *was* expanding; and it *really did* have a beginning.

The discarded image

C. S. Lewis (1898–1963) was an undisputed giant in the world of English literature; his *Chronicles of Narnia* volumes have now sold over 100 million copies worldwide. What is less well known is that he was actually a professional medievalist with a healthy interest in the history of science.

His very last volume, in fact, was on cosmology. Entitled *The Discarded Image*, it features in-depth discussion on how medieval thinkers viewed the universe – and *The Observer* commented that it 'may well come to be seen as Lewis's best book'.

Obviously, Lewis included Copernicus. We already know that *De revolutionibus*, although it was certainly read, did not really bring about a sudden paradigm shift – it was widely viewed as a useful mathematical fiction. A century later, however, when Galileo got on the bandwagon and went all heliocentric, everything went a bit crazy. Why? Lewis gets right to the point:

The real reason why Copernicus raised no ripple and Galileo raised a storm, may well be that whereas the one offered a new supposal about celestial motions, the other insisted on *treating this supposal as fact*. If so, the real revolution consisted not in a new theory of the heavens but in 'a new theory of the nature of theory'.[9]

In other words, it was all very well using a Sun-centred model to get some nice new results – just as long as no one claimed that the Sun *actually was* in the centre. When Galileo did precisely this, and pushed for a moving Earth, he was stating that the Earth really *was* moving. It was this shift that caused all sorts of upset.

Similarly, it was one thing for Friedmann to show that general relativity was mathematically consistent with a dynamic, moving universe, and even for Lemaître to posit a hypothetical beginning (later called the Big Bang), but now Hubble's telescopic results had made their claims *real*. The game had changed. The long-established and all-encompassing scientific image of an everlasting and unmoving cosmos was now being directly challenged. Would it – could it – be discarded?

A cooling-off period

Simon Singh describes the situation in the 1930s as follows: 'Einstein changed his view and supported the Big Bang model. But the majority of scientists continued to believe the traditional model of an eternal static universe.'[10]

Part of the problem – as in the time of Galileo – was that some data was missing and some was wrong. For example, early Big Bang models led to the paradoxical claim that the universe itself was somehow younger than the stars residing in it. This issue was later solved by retaking some measurements – but it provided enough of an obstacle to prevent many from joining Einstein on the Big Bang Bus.

What was really needed was some sort of undeniable mathematical-plus-experimental result that could only be true of a space–time that had begun tiny and then exploded outwards to unimaginable size. Could the expanding universe be caught with a smoking gun? The answer was yes – it would eventually be found holding two of them.

Several key steps in uncovering this new view of reality came from the imagination of the witty and creative nuclear physicist George Gamow (1904–1968). Quite the character, he went on to write brilliant descriptions of both quantum mechanics and general relativity for children by inventing the curious story-world of the ever-befuddled *Mr Tompkins*.[11] This ability to picture the most complex ideas in physics in a playful way no doubt contributed to his role in forcing others to take the Big Bang seriously.

The very early universe, Gamow surmised, would have to be stupendously *hot*. This was because it would contain the same amount of energy as it does now, but this would all be concentrated in a tiny space. In a mini-universe this hot, not even atoms could hold together. Instead, there would be a broiling mass of their three constituent parts: protons, neutrons and the previously mentioned electrons.

As space–time then *grew*, however, it would also inevitably *cool*, since its energy would be getting more and more spread out. Gamow wondered what would happen to this superhot proton–neutron–electron broth as the temperature went down. As we shall see, this question would prove to be more crucial than anyone might ever have guessed.

The small matter of nucleosynthesis

Working with various other scientists over the next decade, Gamow was able to demonstrate that the first major change would be the formation of *nuclei* – this happened when the protons and neutrons collided and stuck together. He and his colleagues showed that the most likely elements to be formed were the simplest: one-proton hydrogen followed by two-proton helium. In fact, they could calculate how much of each would be produced as the universe cooled – and this result agreed with observation.

The building up of nuclei as the temperature decreases is called *nucleosynthesis*. The Big Bang model now had a major piece of evidence on its side – not only could it explain how hydrogen and helium formed, it could say why we see the amount of each that we do. Big Bang Nucleosynthesis is a big deal – and it still holds true today.

Interestingly, one of Gamow's co-workers, Ralph Alpher (1921–2007), calculated that the nucleus-building process would have been completed in just the first 300 seconds of universal time. In April 1948, the *Washington Post* got wind of this thought-provoking result – and responded by running the dramatic headline 'World Began in 5 Minutes'.

Nucleosynthesis was a huge breakthrough, but it was not complete. It could not yet explain the existence of heavier elements and did not, therefore, win over the masses. It did make a second prediction, though – and this one would be the eventual clincher.

Much later into its expansion, when the universe was around 300,000 years old, it would finally be 'cold' enough for the electrons to join the party. Rather than flying around on their own, as they had been up till now, they would bind with the protons and neutrons to make the very first atoms. One extraordinary consequence of all this was that space as a whole would abruptly become *transparent*.

Before this, the sea of roving electrons had essentially prevented the travel of light – but the newly formed atoms would no longer have that effect. At this vital point in universal history, then, a whole cohort of highly distinctive light waves was suddenly loosed to wing its way around the cosmos. And, what's more – if the Big Bang theory was correct, of course – these rays should *still all be out there.*

Digging the cosmological heels in

Meanwhile, back in the 1600s, Galileo's insistence that heliocentrism was true was really not going down well at all. He met protests from many sides: some cited Scripture against him; some questioned his data; some supported Aristotle's thought against any and all opposition; some, quite frankly, just did not like him very much.

He didn't win the argument. In fact, those who adopted the Sun-centred view over the next two centuries did so more because they admired its beauty and relative simplicity than because there was any cast-iron evidence for it. Final 'proofs' of the idea did not actually materialize until the 1800s – by which time most people had changed their minds anyway.

Gamow too faced opposition – in the formidable form of Fred Hoyle (1915–2001), the hard-nosed British astronomer. Hoyle did not like the Big Bang theory at all; it was he who coined the term 'Big Bang' while energetically dismissing the whole thing as claptrap in a radio interview. Realizing that he had to deal with Hubble's data, though, Hoyle and his supporters needed to find a way in which the universe could somehow be expanding, yes – but most certainly not have a beginning.

Their exact motivation for this is not crystal clear, but the moment of creation is one that caused many scientists to revile the Big Bang: it was simply far too indicative of a Creator, which they did not want to be part of the game. Hoyle, who worked closely with colleagues Hermann Bondi (1919–2005) and Tommy Gold (1920–2004) to dethrone the model, hinted on more than one occasion that his dislike of the Friedmann–Lemaître–Gamow model was partly due to its resonance with Christianity.

In its place, then, the triumvirate proposed the *steady state theory*. In this, the galaxies are indeed moving away from each other at high velocity, but new matter is being constantly and spontaneously created at all locations in space to fill in any galaxy-gaps. As a result, the cosmos always looks the same – it *is* unchanging after all. The steady state theory allowed Hoyle to have his cake and eat it. He could acknowledge Hubble's redshift results, but remain loyal to an eternal universe. He had got rid of that blasted beginning.

Others who felt uneasy about the theological implications of a universal 'creation' were thankful for an alternative, as Hawking points out in his *Brief History of Time*:

> Many people do not like the idea that time has a beginning, probably because it smacks of divine intervention . . . there were therefore a number of attempts to avoid the conclusion that there had been a big bang.[12]

Theoretical physicist and Nobel Laureate Steven Weinberg is one such example of Hawking's 'many people'. He makes this clear in a statement that reads rather like a topsy-turvy version of the original Galileo affair: 'The steady state theory is philosophically the most attractive theory because it least resembles the account given in Genesis.'[13]

As it turns out, though, Weinberg was forced to add an extra clause to this phrase – one that laments the eventual triumph of the Big Bang: 'It is a pity that the steady state theory is contradicted by experiment.' For, in a series of new calculations and chance discoveries, Hoyle's last-ditch attempt to hold on to an eternal universe would have to be abandoned. The cosmos *did* have a beginning after all. What was the proof?

Brainwaves and microwaves

What of the missing light that was suddenly released when the coming-of-age universe celebrated its 300,000th birthday? Well, the aforementioned Alpher had determined that, as space expanded and time progressed, these distinctive rays would have been stretched out and lost energy: he was therefore able to predict that the cosmos should be full of a very specific wavelength of electromagnetic waves – *microwaves*.

This cosmic microwave background (CMB) should have been evenly distributed in all directions and locations – so Alpher urged scientists in the 1940s to look for it. They ignored him: some didn't believe his ideas and others thought the CMB would be impossible to detect even if they did try.

It wasn't until 1965 that someone found it – and the irony is that Arno Allen Penzias and Robert Woodrow Wilson did it by accident. Just as with the electron-wave work of Davisson and Germer, the duo were trying to do something else entirely – but they kept being hampered by a constant noise signal through their antenna.

Then, getting wind of the methods of two other American physicists, who *were* interested in the CMB, the truth began to dawn on them. After all, their noise was in the microwave range, it was coming from space and it was equally strong in every direction. Hawking takes up the story:

> Dicke and Peebles were preparing to look for this radiation when Penzias and Wilson heard about their work and realized that they had already found it. For this, Penzias and Wilson were awarded the Nobel prize in 1978 (which seems a bit hard on Dicke and Peebles, not to mention Gamow!)[14]

Be that as it may, Penzias and Wilson had indeed discovered the long-predicted all-encompassing CMB, the indisputable signature of the Big Bang, right across the sky. A universal beginning now was unavoidable: the writing was not just on the wall – it was everywhere else too.

In the meantime, other observations put the final nail in the coffin of the steady state theory. Older galaxies were found to be more common the further away you looked – something expected with the Big Bang model, but not in the constant creation everywhere of Hoyle's proposition. In an interesting twist, Hoyle also did Big Bang physics a favour by brilliantly managing to show that heavier elements could be formed in the centre of large stars – thus solving the problem that Gamow's nucleosynthesis could only make the lighter ones.

Just as Copernicus's heliocentrism had eventually won out after years of battles, everything was now in place for the Big Bang Theory to take its throne. Discovered by a priest, dismissed and then supported by Einstein, ruled out by a dogmatic scientific community, resuscitated by Gamow, attacked by Hoyle, deemed too Christian by Weinberg (and others), proved by the fluke discovery of the CMB – the expanding universe, and its oh-so-unexpected beginning, were here to stay (see Plate 5).

Cosmology and salvation

Although we have been discussing physics, we have never been all that far away from the God question – just as was the case with both gravity and QM. Copernicus and Galileo, as it happens, were themselves both committed Christians and they saw their celestial analysis as part of learning about who God was.

Similarly, C. S. Lewis – the atheist-turned-believer who analysed heliocentrism in *The Discarded Image* – refers once again to our great star when explaining his own world view: 'I believe in Christianity as I believe that the Sun has risen, not only because I see it but because by it I see everything else.'[15]

Owen Gingerich, our Harvard astronomer who was so fascinated by Copernicus, says his study of our universe and its contents have helped him to know better the God of the Bible – and that this trend might continue with future discoveries: 'We can hope that our increased scientific understanding will eventually reveal more to us about God the Creator and Sustainer of the cosmos.'[16]

It is clear, then, that the material in this chapter has resonance with the Big Questions that Hawking posed for us. Yet what of Hawking himself? Does the Big Bang theory hold any significance for him?

Well, first, the phrase 'Big Bang' gets more than one hundred mentions across *A Brief History of Time* and *The Grand Design*. Second, his astonishing 1966 PhD thesis, which began his stratospheric career, is entitled 'Properties of expanding universes'. And, third, much of his most controversial and imaginative physics uses notions involved in the Big Bang theory as its launchpad. In other words, Hawking and the Big Bang are virtually inseparable.

We are nearly at the point, indeed, where we can recount the breakthroughs that proved Hawking to be such a wonderfully inventive scientist. What did he come up with? Why does it matter? How does his work relate to universal expansion; to space–time; to beginnings; to God?

Before we can do that, however, we need to tackle one more issue. For, deep at the heart of the three theories we have studied thus far – general relativity, quantum mechanics and the Big Bang – is a crisis. All are mathematically sound. All are supported by experiment. All enjoy the support of the scientific community in general. Yet all is not well: the tensions between the three are almost unbearable and they threaten to tear cosmology itself to pieces.

In the next chapter, therefore, we will study this threat in all its terrifying glory. Why, exactly, is physics – and the universe with it – in so much trouble? Can it be saved? We shall see that perhaps it can – and that Hawking might just be its Knight in Shining Armour.

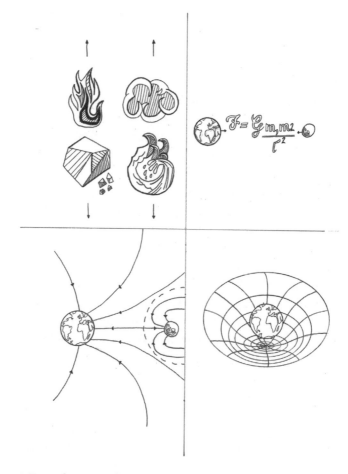

1 **Four theories of gravity**
Top left Aristotle used the four elements to explain why objects fall.
Top right Newton's equation for gravity acting at a distance, but without a physical explanation.
Bottom left Field theory, adapted from Faraday's work on electromagnetism.
Bottom right Einstein's general relativity, in which mass bends space–time.

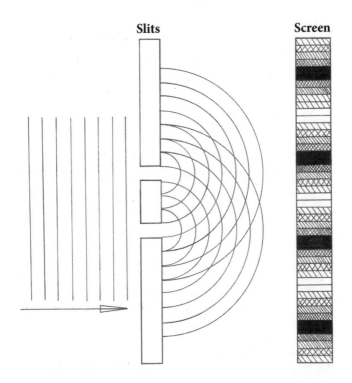

2 Waves double slit

Young's famous double slit experiment. Here, waves arriving at the double slit pass through each one simultaneously, spread out, overlap and form an interference pattern of light and dark fringes on the screen.

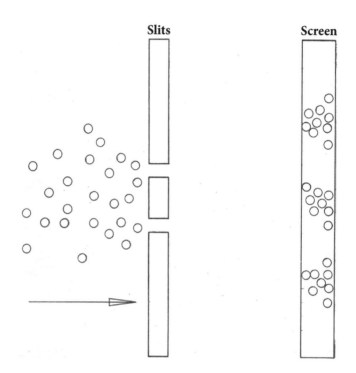

Slits

Screen

3 Electron double slit

The extraordinary electron double slit result. Individual electrons fired at a double slit produce an interference pattern on the screen, as do waves in Young's experiment. This suggests that the electron particles have wavelike properties – evidence for de Broglie's wave–particle duality.

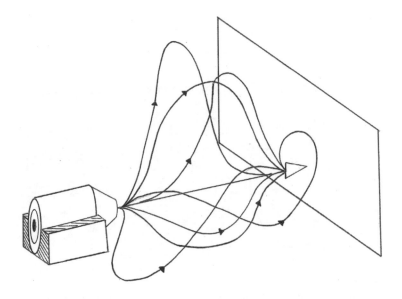

4 Sum over histories

Feynman was prepared to imagine that an electron took multiple different paths (histories) from its origin to a collision point on the screen. Adding up all these paths, he reasoned, would give the probability of the electron arriving at that particular spot.

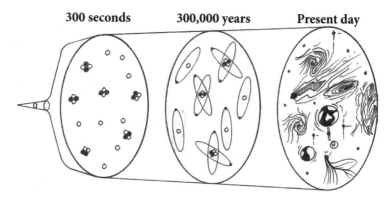

300 seconds   **300,000 years**   **Present day**

5 **Big Bang**

The universe begins from just one point – the singularity that Hawking identified – and then expands outwards. After just five minutes, nuclei have formed. After 300,000 years, electrons are captured and the first atoms come into being. It is at this point that the cosmic microwave background is released. Eventually, stars form and galaxies coalesce. We have reached the present day.

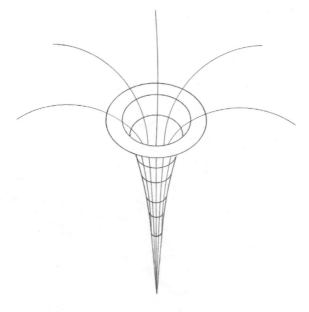

6 Black hole
First discussed by John Michell in classical physics,
black holes re-emerged in general relativity as infinitely
warped space–time. Here, a black hole is characterized by
the space–time grid collapsing down to a single point –
a singularity of infinite density.

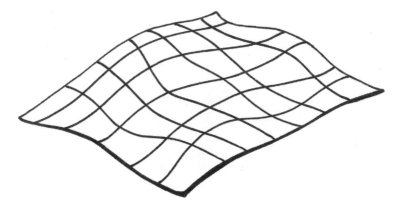

7 **GR background**
The physics of general relativity is played out on a smooth space–time grid. It can fold, bend or twist in the presence of mass or energy – but it *never* breaks.

8 QM background

The physics of quantum mechanics describes a broiling mess of virtual particles that pop in and out of existence. This leads to an ever-changing space–time froth that is torn, has holes in it and is broken down all over the place. This stands in direct contradiction to the background of general relativity.

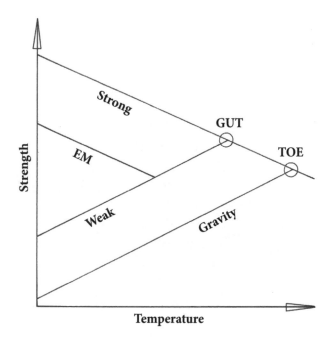

9 Forces

The hope of a theory of everything. There are four fundamental forces in nature: electromagnetism, gravity and the strong and weak nuclear forces. Weinberg and others show that electromagnetism and the weak force can be combined at higher temperatures. If this can be done with the strong force, we have a grand unified theory; if gravity can be brought in, we have a theory of everything. Currently, both possibilities elude us.

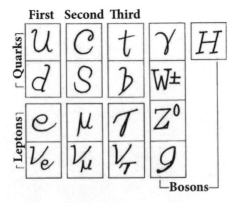

First Second Third

10 Standard model

In the absence of a theory of everything, physicists have grouped the assorted particles in nature into a grid known as the standard model. Despite its neat structure, the strong nuclear force is not really fully inculcated – and gravity is missing altogether.

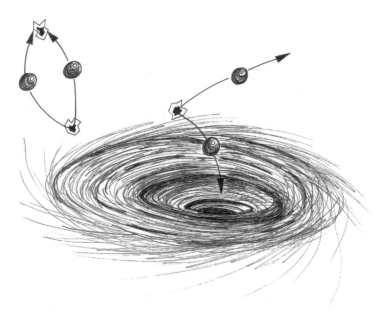

11 Hawking radiation

Virtual particle pairs can pop into existence and then disappear again – this is happening all through space at all times. Near a black hole, however, it is possible for one of the pair to be pulled into the singularity, leaving its partner free to escape. This escapee travels the universe, appearing to have been emitted from the black hole. Particles like this form Hawking radiation.

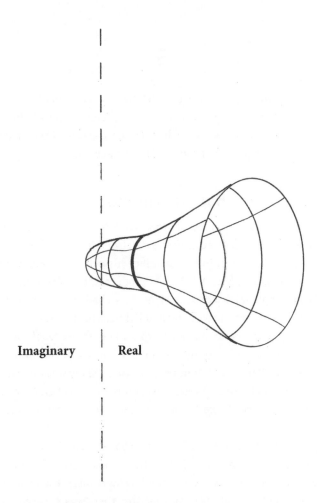

Imaginary | **Real**

12 No-boundary proposal

Hawking and Hartle used the idea of imaginary time to reshape the beginning of the universe. In this imaginary regime, there is no singularity – shown here as a rounded end in which space–time does not collapse to a point. After an arbitrary period, it becomes necessary to switch back to real time, in order for the universe to unfold as we know it.

5

Batman Forever?

On the face of it, *Batman Forever* was a remarkably successful movie. Its opening weekend, in April 1995, saw it bring in $52 million – more than any previous film in US box office history. Many Hollywood bosses were taken aback by this; its predecessor, *Batman Returns,* had been a financial disaster and Tinseltown had all but given up on the Caped Crusader. He was, to put it lightly, a busted flush.

This reboot, however, had a pair of aces up its sleeve. The studio had somehow managed to recruit not one but two of the industry's very best: Jim Carrey and Tommy Lee Jones. Carrey was *the* red-hot name of the mid-to-late 1990s – everything he touched turned to gold. Jones had won the Best Actor Oscar in 1993 and was universally respected as a genius.

The superfamous duo covered quite the cinematic range too – Carrey was a hyperactive and more-than-a-little-crazed screwball, while Jones was a focused and smoulderingly intense screen presence. Getting them into the same picture, then, was a considerable coup. With these two talents on board, there were good grounds for optimism – maybe more folk should have been seeing the (million) dollar signs than did.

Is money, however, the truest measure of a movie? Do idealistic up-and-coming screenwriters dream of quick cash – or do they pine to pen the perfect picture? Do retiring greats reflect on the dosh they have earned – or do they find fulfilment having filmed a fan favourite? Surely, to some at least, it is the art of making something truly generation-defining that matters the most. Did *Batman Forever* – as bountifully rich as it undoubtedly made somebody somewhere – also do well in the I-know-where-I-was-when-I-first-saw-it department? Hardly. It was panned, incessantly, from pretty much all quarters. Critics and punters hated it.

It turned out afterwards that *Batman Forever* had been artistically hamstrung from the start. A hidden (and wholly unexpected) flaw had been sitting right at its heart all along: Carrey and Jones couldn't work together. These unquestionably brilliant silver-screen giants, twin pillars brought in to hold the high-risk project firmly up, were at loggerheads. In fact, it was something of a miracle that the whole darn thing did not come crashing spectacularly down – right on the top of the both of them.

The clash of the titans

In an interview with comedian Norm Macdonald a full 23 years later, Carrey recalled the feud – with a freshness suggesting not much had mellowed with time. The night before their biggest scene together, he recounted, had seen the already worsening matters come to a head.

Carrey had walked into a restaurant near the studio, only to be informed that Jones was already there, dining with some companions. Deciding (probably provocatively) to say a quick hello, the all-energy Canadian headed to the corner his quieter colleague was sat in. Events turned sour almost immediately, as he explains:

> I went over and said 'Hey Tommy, how are you doing?' and the blood just drained from his face . . . he got up, shaking – he must have been in mid kill-me fantasy or something like that. And he went to hug me and he said, 'I hate you. I really don't like you.' And I said, 'What's the problem?' and pulled up a chair,

which probably wasn't smart. And he said, 'I cannot sanction your buffoonery.'[1]

Why, exactly, Jones felt so strongly has never surfaced – but that hasn't stopped people from speculating. Macdonald thinks it was plain old jealousy; Carrey was the new kid on the block and the senior man didn't enjoy sharing the limelight.

Whatever the underlying cause, the total lack of any (positive) chemistry between the two scuppered the film; the detrimental effect of their horrendous relationship destroyed any chance of *Batman Forever* getting on anyone's must-watch-again list.

The director, Joel Schumacher, could perhaps be reprimanded for failing to get his leading men on the same page, but it turned out that he was having his own issues – he described his Batman, Val Kilmer, as 'the most psychologically disturbed human being I have ever worked with'.[2] Maybe the ongoing war between Carrey and Jones had claimed Kilmer as collateral damage.

Despite all this, the movie business – and certainly its bankers – can still call *Batman Forever* a resounding victory. It set box office records and proved that the Dark Knight was a viable character for future pictures, many of which have gone almost interstellar since.

The whole situation, then, can be summed up as follows: *Batman Forever* did what it was pragmatically supposed to do – very successfully, thank you very much – even though it was being simultaneously undermined by the total incompatibility of its two biggest names.

Interestingly, and somewhat bizarrely, the preceding sentence can also be applied, word for word, to something else entirely: the last one hundred years of theoretical and experimental physics.

Why can't you two just get along?!

In Chapter 2, we saw how Einstein finally managed to crack the problem of gravity by reimagining space and time. He pictured them as

82

being woven into a single bendable sheet called *space–time*, which is warped by the presence of mass – or, as it happens, of *energy*, which will be important later in this chapter.

The resulting theory, *general relativity*, was developed in 1916 and it still holds true today. Time and again, its predictions have been put to the test and experiments have proved them right on every single occasion. General relativity or GR beautifully describes our universe on the grandest of scales. In scientific terms, it is a bit of a superstar.

Chapter 3 then introduced the weird world – stumbled across in 1905, by Einstein again – of quantum mechanics. Subatomic reality, it turned out, had been hiding a radical new rule-set – equations so novel that all those outside them were termed 'classical', even if they were from the twentieth century themselves.

In QM, our common sense is not of great use – the behaviour of the very small is not like anything we ever see in our everyday lives. Particles can be waves; waves can be particles. A single electron can go through two slits at the same time. The moon might only be there when we look at it. The uncertainty principle denies us pure knowledge: we can know one thing or another, but never both. And yet, just like GR, QM has aced every laboratory test we have thrown at it.

The result of these two wonderful discoveries has been unparalleled scientific success. Between them, GR and QM have given us Apollo 11, the Big Bang and computers – quite the ABC. And, if that wasn't already enough, they combine to give us the sat nav, which depends on QM for its circuitry and GR for its signal. We would, literally, be lost without them.

Pragmatically, then, we can only conclude that all has gone really rather well – if *Twentieth-Century Physics* was a movie, it would have smashed every box office record in town. GR has us covered with the supersized stuff; QM takes over when the lumps get little. Just like Carrey and Jones, they boast differing skill sets and impeccable track records – but that's not all they have in common with that duo.

If anything, things are even worse in modern physics than they ever were in the poisonous atmosphere of *Batman Forever*. At least Carrey and Jones both finished filming without one bumping the other off; the battle between QM and GR will not end so nicely. One or both of them will have to go – the clash is a fundamental one, written in the stars (or the atoms, depending on the victor). There is no reconciling the two as they currently stand. The aim of the rest of this chapter is to explain *why*. Val Kilmer, beware: this is not going to be good for your health.

Separation anxiety

How can it be that physics has landed men on the moon, found ways to image the workings of our brains and given us high-definition always-on live streaming from a webcam at Lanzarote Airport,[3] if it is, at heart, governed by two theories that despise each other?

The first answer is that, for most intents and purposes, it actually doesn't really matter all that much if QM and GR don't agree. They don't tend to devour each other for the same reason that polar bears don't tend to eat penguins – they are incredibly unlikely to meet. The brilliantly titled *The Complete Idiot's Guide to String Theory* puts it this way: 'Ordinarily, when gravity is strong, quantum effects are weak, and when quantum effects are strong, gravity isn't.'[4] In other words, physicists generally need to use one or the other for a given problem, but not both. Their realms are so different that they can accurately be described – just like the bears and the penguins – as being polar opposites. Why?

Well, for starters, gravity is an incredibly weak force. This is a fact that is not immediately apparent, but it is easier to recognize once we realize that when we lift an apple, we are overcoming the gravitational pull on it *from the entire Earth*.

In fact, gravity gets rather pathetic as soon as the masses involved are much smaller than planets – the gravitational force between two apples, for example, is so little that there is simply no point in ever considering it. We only really need GR in strong gravitational fields – to track comets or stars or galaxies.

QM, on the other hand, deals with the three other forces in our universe: *electromagnetism*, the *weak nuclear force* and the *strong nuclear force*. All three of these are far, far more powerful than gravity – even the least of them (the weak force, shockingly) is one-hundred-thousand-billion-billion-billion times stronger than Newton's brainchild. Because of this, QM doesn't need to involve GR in its calculations – the difference that including gravity would make is so vanishingly small it would be an utter waste of time.

If gravity is so spectacularly weedy, though, why does it *ever* matter? Why isn't it brushed aside by the three burly forces of QM every single time? The answer is a matter of *scale*. Two of QM's three forces, it turns out, are extremely short-ranged: the strong force can just about stretch across a small nucleus and the weak force only manages a tiny fraction of that. Anything atom-sized or bigger is well beyond their reach, meaning they can be totally ignored.

The third force, electromagnetism, is not so encumbered: it has an *infinite* range. And yet, despite this limitless reach and its herculean strength, it still loses out to gravity when things get big. How can this possibly be?

The answer is surprisingly simple; it derives from the fact that electromagnetism is the force between *charges*, while gravity is the force between *masses*. Crucially, our universe contains two types of charge – *positive* and *negative* – but only one type of mass (positive). This means that electromagnetism both pushes *and* pulls its victims, while gravity *only ever pulls*.

This is hugely significant. Any reasonably sized object contains an equal mixture of positive and negative charge, meaning that the mighty electromagnetic force actually defeats itself by pushing and pulling it at the same time, cancelling itself out. The mass of the object, on the other hand, is purely positive. This means that gravity, with stubborn persistence, just keeps on pulling, and pulling, and pulling. Even though each individual pull is puny, it all adds up in the end: gravity wins.

In short, gravity is the only force to achieve anything over large distances – the three others, potent as they are, disappear completely. Drop down to nuclear scales, though, and the reverse applies – the triumvirate run the show with their prodigious power, totally dominating their scrawny cousin. All this gives physicists a simple guide as to which set of rules to use for their particular domain. Gauging galaxies? GR. Evaluating electrons? QM. Simple.

This is all rather pleasing. It looks, reassuringly, like any deadly conflict between the two will for ever be avoided. Science can be allowed to continue, unabated, on its wonderful victory parade. Any previous anxiety about the clash can be dismissed, since the polar bears of GR and penguins of QM are more than sufficiently separated by the vastness of our cosmic landscape. Or, at least, they *would* be – if it wasn't for those pesky *zoos*.

The dark star

If an astronaut were to stand on a small asteroid – say one the size of a town – he would find himself in the gravity-run realm of GR. The maths tells us that he and the rock would be attracted to each other, but not all that much. The attraction is just about big enough to hold on to an insect, but nowhere near what is needed to keep a spaceman on the surface. Any half-hearted attempt at a jump would launch him dramatically off into space – he can escape the asteroid's attentions with ease.

His sister, on the other hand, is exploring a different astronomical body – one far larger. She is on the dwarf planet Ceres, which is roughly the size of the United Kingdom. The gravitational pull is starting to add up now, and she feels more force than her brother. When she jumps, she can get up to quite a height, but is always pulled back to the ground. To get off – and stay off – Ceres is no small matter: she would have to launch herself upwards at an unachievable 1,100 miles per hour.

The speed required to leave a body permanently is known as its *escape velocity*, and it increases with the mass of the body. It turns out to be exactly equal to a second speed: the one with which a far-off object falling towards that same body would strike its surface. Fortunately, this calculation is straightforward and has been understood for centuries – it is carried out using Newton's well-worn laws of motion. Easy.

In 1783, however, Reverend and geologist John Michell (1724–1793) spotted an extraordinary consequence of all this that everyone else had missed. Something very strange would happen, he pointed out, if the mass under question got big. As in *really, really* big:

> If the semi-diameter of a sphere of the same density as the Sun were to exceed that of the Sun in the proportion of 500 to 1, a body falling from an infinite height towards it, would have acquired at its surface *greater velocity than that of light.*

Michell knew that escape velocities should, in theory, keep getting faster and faster as stars got bigger and bigger. Eventually, for a large enough star, they would be quicker than light speed – that led to a genuinely astonishing conclusion: 'all light emitted from such a body *would be made to return towards it* by its own proper gravity'.[5]

In other words, the gravitational pull of this gargantuan stellar mass would be so strong that *even its own light could never escape it*. Despite the fact that it would burn just as bright as (maybe even brighter than) any other star, it would be completely invisible to us. The imaginative rock-collecting vicar had unearthed perhaps the weirdest of all the artefacts found in our increasingly bonkers universe: the brain-befuddling *black hole*.

Where QM and GR meet

If Carrey had chosen a different restaurant that fateful night – or if he had opted for a takeaway – perhaps everything would have played out differently. Jones might have enjoyed a quiet meal with his pals, forgotten his frustrations with his colleague and come back on set the next day refreshed and re-energized.

The two might have gone on to create some magic in their big scene; they might have become good friends; they might even have collaborated further, harnessing their diverse brilliance to great effect. As it was, they did meet and it went horribly. The relationship broke down completely. It has been more than two decades and both men still act, but they have never worked together again.

If QM is Carrey and GR is Jones, then black holes are that table in that corner of that restaurant. If GR is polar bears and QM is penguins, then black holes are just like a (bad and dangerously incompetent) zoo. We said earlier that GR and QM don't generally meet – one is strong when the other is weak – but the rules go out of the window whenever and wherever black holes are concerned. As might be expected, all hell then breaks loose. And hell, for a physicist, is *infinity*.

While mathematicians are perfectly happy playing around with infinities (indeed, they sometimes appear to revel in it), physicists feel very differently. For them, having an infinite quantity turn up in a solution is a veritable nightmare, because they do not live in the abstract – theirs are real claims about the real world and infinities just don't make sense in reality. In an interview with the BBC, astrophysicist Andrew Pontzen put it this way:

> In physics, infinity is really 'bleep' annoying. It's like it contaminates your equations. If you have an infinity lurking somewhere in space it actually messes up our ability to predict the future . . . as soon as there is an infinity in the universe it means that predictions start to break down.[6]

His view is pretty much the consensus. The appearance of infinity in a physics formula is generally interpreted as a sure-fire sign that something very important has gone very wrong – that the theory needs to be fixed, and quickly.

No one seems to have mentioned this principle, however, to the ever-enigmatic black hole. Bewilderingly, black holes are quite content to ignore the feelings of physicists entirely and *actually exist* in our universe – despite the fact that they have infinity coursing, entirely unchecked, through their veins.

This unsettling truth was arrived at in stages. The original black hole that Michell described was based on Newtonian gravity and followed his brilliant-yet-now-superseded equations to the letter. Incredibly, though, just one year after Einstein overhauled Newton's ideas completely, the indomitable black hole popped up again in new-and-improved GR-ready form. This time, it was an interesting and unavoidable quirk in Einstein's new formulae: a super-dense mass that folded space–time in on itself.

The extreme curvature around this odd creature (that had to be several times more massive than our Sun) was great enough to both guide light in towards it and also prevent it from escaping again – its space–time slopes were too steep for the light to climb out (see Plate 6). Just like Michell's version, it would be invisible. Black holes had faced Einstein's remodelling of gravity and survived to tell the tale. In fact, as time rolled on, it would become clear they had much more to say – and some of it was rather disconcerting.

Soon, astrophysicists investigating large stars under general relativity realized that the strong gravity could, under specific conditions, cause a 'collapse'. Precisely what this meant was not clear and the final state of a collapsed star was uncertain – but one man, now known for a different reason entirely, was intrigued by the notion. Taking the maths further than anyone else was prepared to, he wrote what Werner Israel (an expert in GR and collaborator with Hawking) later referred to as 'the most daring and uncannily prophetic paper ever published in this field'.[7]

This far-thinking man was none other than J. Robert Oppenheimer (1904–1967), of Manhattan Project and nuclear bomb fame. When analysing collapsed stars, he wondered what would happen if no artificial limit was placed on the process. Ignoring the various different stopping-points that others had built into prior models, he decided to let the star keep on going.

It obliged. As it shrank, it retained all its starting mass, but this was contained in an ever-decreasing volume, making the star denser and denser and denser. This caused the space around it to bend even more, which pulled the shrinking star in even harder, which increased the density and curvature – and so on. Science writer Marcia Bartusiak completes the terrifying story:

> [Oppenheimer and a colleague] calculated that the star would continue to contract indefinitely. There is no rest for the weary when gravity takes over. The matter within such a collapsing star is in a state of permanent free fall . . . Space-time is so warped around the collapsed star that it literally closes itself off from the rest of the universe.[8]

And herein lies the problem. First, the massive star is governed by its overwhelmingly large gravity, which puts us squarely in the kingdom of GR. Second, it eventually becomes ridiculously small – which is, of course, the territory of QM. Third, we hit a host of banned infinities; that phrase 'ridiculously small' from the previous sentence, it turns out, is very much an understatement. Bartusiak again:

> They figured that the star collapsed to a point, a singularity squeezed to infinite density and zero volume (which seems impossible). Their equations indicated this, but they hesitated

to say it directly. That's because singularities are a horror to physicists.

The cause of the horror is that long-established tell-tale sign of a theory in crisis. For, in the middle of a collapsed star – a black hole – is an infinitely small, infinitely dense and infinitely curved *singularity*.

This, then, is what happens when GR and QM finally meet: unmitigated theoretical disaster. At least Carrey and Jones managed not to come to blows. Here, in the world/restaurant/zoo of QM plus GR – of *quantum gravity*, as it is known – there is blood on the walls. Lots. An *infinite* amount, in fact.

Hawking destroys the universe

So far, the only real problem we have with QM and GR is that they meet in black holes and that the result is infinities that actually exist. Can this truly be called a catastrophe? After all, black holes are *weird*: they are 'out there' and otherworldly; they feel far more sci-fi than sci-fact. Who cares if infinities show up in black holes? Aren't we in danger of overstating the case a little? The answer is yes, we probably would be – if it wasn't for a certain Stephen Hawking and his 1966 PhD thesis.

The recently diagnosed Hawking could well have been forgiven for abandoning his work as a young postgraduate – there were no guarantees that he would even finish his doctorate before his disease got the better of him. Instead, he ploughed on and completed it; it is now the most read doctoral paper in history. So what did it actually *say*?

To understand the answer to that question, we will need a little update. Ever since black holes had appeared in GR, physicists had felt sure that they could not be real. Even if they were, it was suspected, they would most certainly not turn out to have singularities

91

at their centres. The widely held belief was that some hitherto unknown physical principle would one day be discovered that would prevent the total gravitational collapse of a star – and protect the universe from all those nasty infinities.

Among these believers was Einstein himself. In 1939 he decided to take matters into his own hands and prove, once and for all, that there could never be a black hole (at that time, still called 'Schwarzschild singularities' after their GR discoverer) in nature. After pages and pages of complicated calculations, his paper drew to a close with the words:

> The essential result of this investigation is a clear understanding as to why the 'Schwarzschild singularities' do not exist in physical reality . . . [they do not] appear for the reason that matter cannot be concentrated arbitrarily. And this is due to the fact that otherwise the constituting particles would reach the velocity of light.[9]

The great man thought he had resolved everything – but he hadn't. His desperation to remove singularities and restore predictability to all physics led him to work with invalid assumptions; Bartusiak declares this to be Einstein's 'worst scientific paper'.

Over the next twenty years, the landscape changed remarkably. Cosmologists, picking away at the black hole issue, kept stumbling across infinities that didn't disappear easily; but they kept faith that a deep truth would emerge and destroy them. Then, in 1965, a mathematician came along – and put an end to these physicists' hopes.

Roger Penrose, at home with abstract reasoning, saw the whole thing quite differently. To him, the singularity at the core of a black hole seemed both natural and mathematically vital. In a paper just one fifth of the length of Einstein's clumsy dismissal of them, Penrose showed singularities were inevitable. It didn't matter what shape the star was or what it was made of – there was no way out.

The infinities were real. The debate was over. Physics had to let infinity in, even if its name wasn't on the list. The black hole argument was settled.

It had already been established that the equations of GR broke down completely at a singularity, meaning all its predictive power would be lost. Now, thanks to Penrose, singularities were unavoidable. This meant that, in certain small pockets of the universe, all bets were off – no one could do any science at all in these infuriatingly infinite regimes. Still, at least it was only happening at the centre of the already deeply mysterious black hole. Where's the great harm in that?

Enter Hawking and his deadly dissertation. Hot on the heels of Penrose, he dealt a devastating blow to our conception of our world and its knowability. In a stroke of magisterial genius, he realized that he could combine Penrose's analysis of black holes with general relativity's description of the Big Bang. Thinking bigger than anyone else in cosmology, this young researcher joined up the dots. The picture that emerged was a stunning one: it was elegant and it was terrifying.

Like many great ideas, Hawking's doctorate is surprisingly simple. The Big Bang model of GR (see Chapter 4) speaks of a universe that began very *small*. It then expanded, dramatically, *outwards*. Penrose's black holes, on the other hand, start off *big* – then GR pulls them, dramatically, *inwards*. Hawking's incredible insight was to realize that the two processes are *mathematically identical* – one is merely the reverse of the other.

Run time backwards on our cosmos, he could show, and it would behave exactly like one of Penrose's shrinking suns. He took maths originally intended for single stars and applied it to nothing short of everything. This led to a truly staggering result: our universe – the whole thing, every part of it, the complete show, *space in its entirety* – was once nothing more than a singularity. In typically straightforward style, he states in the introduction that his work: 'deals with the occurrence of singularities in cosmological models. It is shown that a singularity is inevitable provided that certain very general conditions are met.'[10]

It is hard to overstate the significance of this finding. By claiming that the whole of reality was once bound up in an infinitely dense, infinitely small single point, Hawking had placed the beginning of our existence firmly out of our reach. No one, now, could know where we had come from. No one could predict how the universe would unfold – physics broke down right at the start. Fundamental truths about why the cosmos is the way it is were suddenly hidden from view behind an infinite curtain.

This earth-shattering, science-breaking finding came in the fourth chapter (of four) of a rather nondescript paper with a rather nondescript title, written by a relatively unknown and terminally ill 24-year-old. And yet, cosmologically at least, it is no exaggeration whatsoever to say that Stephen Hawking's 'Properties of expanding universes' changed everything. He had arrived with a bang. A Big Bang.

QM versus GR – the infinite battleground

In proving that the universe itself began with a singularity, Hawking had done something extraordinary – and its consequences are still felt today. The beginning of the cosmos was rendered unscience-able and even second prize now looked unwinnable. For, after the initial and inscrutable singularity, came the next rather troubling stage: a very, very small universe with lots and lots of mass.

The 'small' part quite clearly meant QM. The 'lots of mass' part demanded GR. There simply was no other way round it – the early universe required them both. Any lingering hope that the quarrelling cousins could be kept apart for ever were done for. Sure, plenty of physics could still be achieved in other areas – and it was. The Big

Bang and the origin of the universe, though? Well, that was going to need a new theory: one that combined QM and GR. And no one knew how to do that.

To see why, it is probably worth mentioning a few of the other problems that appear when the terrible twosome meet. Whole books have been written on this topic, so we can only skim the surface – but even the surface shows clear signs of battle.

Skirmish one: holes

The uncertainty principle (see Chapter 3) is the heartbeat of quantum mechanics. This principle, discovered by Heisenberg, puts limits on what we can ever know for sure. Pairs of measurements inhibit each other: the *more* we know about the amount of *energy* a particle has, for instance, the *less* we can know about the *time* it will have that energy for.

This doesn't matter when we consider large-scale phenomena such as our own bodies, but at the level of single particles it causes havoc. Incredibly, it tells us that particles with small amounts of energy can pop into existence out of 'nothing' – provided they only exist for a very, very short period of time. The equations of QM allow this, but particles must appear in equal and opposite pairs and then disappear again, into nothingness, straight away. We call these *virtual particles*.

This is happening all the time, all through space. The submicroscopic universe is dominated by a seething, broiling mess of these ghostlike virtual particles appearing and disappearing, over and over again – countless numbers of them, never stopping, never slowing. Do the maths and this incessant chaotic dance causes space–time, on the finest of scales, to fold in ways it isn't supposed to. The rapid and random energy changes pull it in directions it cannot handle – it rips and tears; it fills up with holes; it becomes unrecognizable.

General relativity, on the other hand, knows none of this. It was formulated on the very notion that space–time is smooth and well behaved: there are no rips allowed. Include the holes caused by QM, then, and GR just simply stops working. Try to calculate a measurement this way, says theoretician Brian Greene, and you'll get 'the same ridiculous answer: infinity'. He continues:

> Like a sharp rap on the wrist from an old-time school teacher, an infinite answer is nature's way of telling us that we are doing something that is quite wrong. The equations of general relativity cannot handle the roiling frenzy of quantum foam.[11]

In short, general relativity asks for a smooth and unbroken backdrop, but quantum mechanics supplies one ravaged with holes (see Plates 7 and 8). GR simply cannot operate on a QM footing. Bizarrely, the opposite is also the case.

Skirmish two: shifting sands

Quantum mechanics does its best work when it is combined with the field ideas first introduced by Faraday (see Chapter 2). The robust and powerful outcome – *quantum field theory* or QFT – is superb at predicting the behaviour of just about any event that might ever occur in the subatomic realm. Just as with Faraday and Maxwell's models, these predictions come from studying the behaviour of the fields involved – how they move, change or vibrate in space–time.

The killer question, though, is *which* space–time? The fixed, static, passive space and time of classical physics, or the bendable, dynamic, relativistic space and time of Einstein's gravity? The surprising answer is that QFT is built on the former, not the latter. As far as the geometry of space–time is concerned, QM and QFT operate as if gravity does not exist.

Lee Smolin, a man who has spent his career looking for ways to make peace between QM and GR, bemoans the decision made by earlier quantum physicists to construct their theory on a non-relativistic background. He believes that their refusal to develop QM on the moveable background of GR has cost us all dearly ever since:

> Ignoring gravity meant taking a step backward, to the understanding of space and time before Einstein's general theory of relativity. This was a dangerous thing to do, as it meant working with ideas that had been superseded . . . the chief lesson from general relativity was that there is no fixed-background geometry for space and time.[12]

Later on, Smolin reminds us of the consequences of this controversial call: 'you have to be very careful not to get infinite or inconsistent answers'. QM gives GR an unusable foundation, and GR returns the (unwelcome) favour. These are not minor concerns – they are, by very definition, fundamental.

Skirmish three – black holes everywhere

Fundamental differences in the basics are bad enough; how could it get worse than that? Well, the QM–GR rift isn't just foundational, it threatens to stop us knowing anything at all. For, rather worryingly, all we have to do to cause further infinite chaos is ask *where* or *when* something happened.

Consider a single particle – an electron, say. If we want to know *exactly* where it is, we can't – the quantum mechanical uncertainty principle doesn't let us. It does, however, offer us a deal: we can pin the electron down with greater and greater accuracy if we are prepared to use more and more energy to make the measurement. Without GR in the picture, this is weird, but still workable – which means it is just about acceptable.

Include GR, though, and we hit a disaster. To figure out where the electron is to the precision needed for gravitational calculations, the energy we have to use is vast. It is so vast, in fact, that GR tells us the space–time around the particle we are investigating will be bent infinitely out of shape: we end up with a black hole.

The same applies to asking *when* an event in space–time occurred. GR needs to know the answer almost perfectly; QM says that will cost big energy; GR responds by forming a black hole. Fantastic. With both on board we can't even know what time it is or where we are. Even the act of asking triggers theoretical infinities. No wonder physicists work with just one of the theories or the other: together, they are a singular nightmare.

Hawking the hero

Is there any hope, then? Can the two protagonists of *Twentieth-Century Physics* be reconciled? Unlike Carrey and Jones, it seems they can't even manage one scene together, let alone ever become partners. Should we abandon the quest for a combined theory and give up entirely on quantum gravity? It is fair, of course, to ask if such a theory is truly *needed* – something that Sean Carroll has done in the past:

> This is obviously a crucially important problem, but one that has traditionally been a sidelight in the world of theoretical physics. For one thing, coming up with good models of quantum gravity has turned out to be extremely difficult; for another, the weakness of gravity implies that quantum effects don't become important in any realistic experiment. There is a severe conceptual divide [between GR and QM], but as a practical matter there is no pressing empirical question that one or the other of them cannot answer.[13]

Carroll's 'extremely difficult' is an understatement. Bringing the two major theories of physics together is not just tough; it has proved – thus far – to be impossible. One might think the problem is recent and that a little more time will see it cracked, but the first attempts to unite Einstein and Heisenberg came nearly a century ago. Physicists are still asking themselves how to do it. It is not clear they are that much the wiser.

And this is where Hawking comes in. With his PhD, he threw a cosmological spanner in the works: he took the QM–GR fight and placed it squarely at the start of our universe. He proved we began with a singularity – an infinity that removes all predictive power. In doing so, he sounded the death knell for the final goal of all science – we would never know why the world looks like it does.

The only hope left was quantum gravity. A correct theory of this type could handle singularities – or perhaps tell us singularities don't exist. It could draw together QM and GR. But *how* do we go about doing this when even asking for the time makes black holes? For decades, the greatest minds in the business have been mystified.

Smolin's bestselling book – *The Trouble with Physics* – is a detailed account of just how hard it has always been to work on a combined theory. Even our best guesses, he admits, are total long shots. He is a man who has experienced the highs and lows of trying to solve the greatest theoretical conundrum of the last one hundred years; his life has been one of shooing singularities and banishing black holes. Yet he chooses, right at the start of his tome, to mention a different question: 'There may or may not be a God.'

For the umpteenth time, then, we see *God* make an appearance in a mainstream book about cosmology – this time on page one, paragraph one, line one. By now, though, we should not be surprised. After all, the connection has been there for millennia – in 700 BC the prophet Isaiah said: 'Lift up your eyes and look to the heavens: Who created all these?' In our next chapter we shall see how Hawking turned hero – partly by following Isaiah's advice. Having declared the Big Bang to be for ever hidden from scientific view, he lifted his eyes once again.

This time, when he looked to the heavens, he saw something new: that black holes weren't so black after all. To the astonishment of his peers, Hawking put a portion of predictability back on the table. For, despite their clear dislike of one another, he showed that QM and GR – just like Carrey and Jones – *could*, under certain circumstances, be made to work together.

Perhaps, he dared to suggest, there *was* some hope after all.

6

Part 1

Akira Yoshida • Unification • The grand unified theory • Putting the heat on • Just how far can we take this? • A new hope • Black holes ain't so black • Hawking radiation • Securing a legacy • Trouble in Paradise

Akira Yoshida

Japanese prodigy Akira Yoshida burst on to the comic book-writing scene in 2004 with a background, skillset, energy and novelty that US editors had previously only dreamt of. Appearing out of virtually nowhere (a few manga pieces at home and some low-profile work in the States) the market-leading *Marvel Comics* snapped him up almost immediately. Yoshida fulfilled a brief everyone was desperately looking for: someone who understood Japanese culture from the inside but could write well for an American audience. And boy, could this guy write.

Within just a couple of months, he had penned major storylines for many of Marvel's biggest franchises: *Thor, Fantastic Four, Wolverine* and even the all-conquering *X-Men*. His star was ever-rising and Marvel were delighted. They had taken a risk with this previously unknown and unproven artist, but it was paying off. Yoshida might even have gone on – under different circumstances – to be the *best*.

So great was this young man's impact that in March 2005 the celebrated *Comic Book Resources* website ran an exclusive feature-length interview with him. In it, Yoshida name-dropped multiple legends of the industry, thanking each for their faith and advice along the way. Somehow, in the space of less than two years, he had managed to work alongside almost everyone worth knowing in the business. His gushing interviewer finished the piece – entitled 'Akira

Yoshida: A bullet for Marvel's young guns' – full of admiration, respect and hope: 'While Yoshida's current plate is full of projects, he shows no sign of slowing down in the near future . . . he has another Marvel series in the works plus a new series for "another major US publisher"'.[1]

Yoshida's rise from obscurity to superstardom was unparalleled. The same, though, can be said for his fall. Suddenly, in 2006, Yoshida disappeared – completely. No one has heard from him since.

At around the same time, a second talented author was also making his way to the top – albeit more slowly. Beginning his career as a translator, C. B. Cebulski was prepared to play the long game. The Connecticut-born penman took on a variety of unspectacular editorial roles at Marvel and peaked during his first period there in 2006, when he was head writer for the *Marvel Ultimate Alliance* video game.

After a short freelance spell Marvel hired Cebulski back. His consistency and intelligent use of social media saw him rise through the ranks and, in November 2017, after nearly two decades of slog, it all paid off: he was appointed Editor-in-Chief. Slow and steady had won the race – Akira Yoshida was nowhere; C. B. Cebulski was heading up the most successful comic book company on the planet.

Despite being in the same trade, the two men boasted radically different stories – in that sense, they could not really be much further apart. Yet strangely, if we turn up the investigative temperature a little, the pair gradually become more alike. In fact, if we let it get red hot, something truly extraordinary occurs: Cebulski and Yoshida are no longer just *similar* – they turn out to be one and the *same*.

Unification

Physics, just like any other enterprise, needs both its Big Thinkers and its Handle Turners. Many of the latter work in one specialized area, dedicating an entire career to nudging that subsection of the discipline along a little bit. There is no shame in this whatsoever; indeed it is vital if progress is to be made. Details matter and those taking smaller, slower steps are less likely to miss them. Handle Turners are far more than just cogs in a machine – they are the ones making sure the cogs actually *fit*.

The Big Thinkers, on the other hand, are few and far between. We have met several of them during our journey – Aristotle, Newton, Faraday, Einstein – their names usually become extremely well known. These heroes tend to have one thing in common: they take ideas that were thought to be separate and find a new way to *join them together*.

Take Faraday (see Chapter 2), for instance. Electricity and magnetism had been studied as two different topics, but he viewed them instead as a package. From these ideas sprang a completely new science, that of *electromagnetism*. He (along with Maxwell, who did the maths) had *unified* the previously distinct realms. Others have filled in the small print, of course; there are tens of thousands turning electromagnetic handles even now. They owe their jobs, however, to Faraday and Maxwell – two of the Godfathers of *unification*.

This pattern has continued ever since. The vast majority of physicists work on small-step iterative projects, while only a handful pursue unificatory breakthroughs. It should come as no surprise to the reader that Hawking was firmly in the second camp – what he yearned for, more than anything, was a theory of *everything*. This goal was not Hawking's uniquely; folk had been targeting it for decades before he turned up. So where, exactly, had they managed to get to?

The grand unified theory

In *The Grand Design*, Hawking eulogizes his fellow Big Thinker Maxwell for his 'set of equations describing both *electric and magnetic forces* as manifestations of the *same physical entity*, the electromagnetic field'.[2]

The end goal in physics is to do the same for *all* forces. We mentioned in Chapter 5 that there are *four* of these – *gravity, electromagnetism, strong nuclear* and *weak nuclear* – and that between them they describe everything in the universe. To unify these four, then, would be the completion (or maybe beginning) of physics. All would be part of the 'same physical entity' – we would finally have a theory of everything (see Plate 9).

As one might expect, this endeavour has turned out to be rather difficult. Hawking gives us just one of the many reasons why:

> Though they both revolutionized physics, Maxwell's theory of electromagnetism and Einstein's theory of gravity – general relativity – are both, like Newton's own physics, *classical* theories . . . if we are seeking a fundamental understanding of nature, it would not be consistent if some of the laws were *quantum* while others were *classical*.
>
> We therefore have to find *quantum versions* of all of the laws of nature.[3]

By 'laws of nature' Hawking means the four forces; we need to quantum-ify or *quantize* each of them. Perhaps surprisingly, early progress on this daunting task was actually quite good.

A quantum picture of electromagnetism was the first to arrive. Dubbed *quantum electrodynamics* or QED, it was figured out in the 1940s by – among others – Richard Feynman. One of the key ideas was the introduction of a new type of particle, called a *boson*, that 'carried' the electromagnetic force with it as it travelled through space.

QED, though, could hardly be called easy. Its key calculations resulted in infinities, those harbingers of physical theory doom. Fortunately, as Hawking explains, the infinities of QED could be gently ushered away: '[QED] seems to imply that the electron has an infinite mass and charge. This is absurd, because we can measure the mass and charge and they are finite. To deal with these infinities, a procedure called renormalization was developed.'[4]

Renormalization was a massive breakthrough – it allowed physicists, in certain specific situations, to get rid of infinite answers. The basic idea was this: introduce some carefully chosen *negative infinities* to almost completely cancel out the positive ones given by the theory, leaving behind only tiny numerical morsels. These morsels will be the exact values of mass or charge or whatever – values already known to be true from experiment.

The dodgy thing about all this is that renormalization can be fiddled with until it gives us the answer we want: the 'result' is therefore a truism. Tune it to get a different and *incorrect* number for the charge of the electron, and renormalization will happily oblige. Hawking is characteristically up front about all this: 'These manipulations might sound like the sort of things that get you a flunking grade on a school maths exam, and renormalization is indeed, as it sounds, mathematically dubious.'[5]

Nevertheless, if the *correct* values for mass and charge are inserted into QED by hand, everything else it then says is spot on – the theory works brilliantly. It is a genuine quantum version of electromagnetism and forms the theoretical basis of modern information technology. This means that one of the four forces is successfully quantized – and we're still only in the 1940s. There are three more still to go: the strong nuclear force, the weak nuclear force and gravity.

That means it is now time – obviously – to revisit Akira Yoshida and C. B. Cebulski.

Putting the heat on

In December 2005 comic book blogger Brian Cronin launched an 'Urban Legend' investigation that began with the following insight: 'Whenever a new creator comes out of seemingly nowhere, people are bound to be curious about them, especially when, in the case of writer Akira Yoshida, the new writer gets such "plum" assignments as the X-Men/Fantastic Four crossover.'[6]

Wondering whether or not the Yoshida phenomenon was totally legitimate, Cronin decided he'd contact Marvel directly to get details. Rather suspiciously, none of the editors said they had met or even seen the Japanese writer – until, that is, he heard from Mike Marts. An experienced pro at both Marvel and DC Comics, Marts proceeded to allay Cronin's concerns: 'You bet – I've had lunch with the guy – very nice guy . . . there's ONE conspiracy theory down the drains!!!'

Less than a year later, Yoshida went off the grid and the story, like him, disappeared. More than a decade later, though – at around the same time as the grand promotion of Cebulski – it came back. And this time, the heat was very much on.

The first to stoke the flames, 13 years after Yoshida had vanished, was former editor Gregg Schiegel. He released a short story as a podcast that told of a Marvel editor adopting an alternative identity and writing for his own publications – a practice that was an industry taboo, against official policy and an instantly sackable offence. To protect himself, Schiegel claimed the podcast was pure fiction; he named all the main players after characters from the political drama *The West Wing*.

Needless to say, this set many tongues wagging – but journalists who formally approached the bigwigs at Marvel were told, quite firmly, that Akira Yoshida *was* real; the so-called 'conspiracy' was a total non-story. This lowered the temperature a little, but not by enough – and not for long enough. Just days after Cebulski's appointment as Editor-in-Chief, a rival in the industry tweeted: 'Hey comics journo friends: we should definitely be asking Marvel and new EiC CB Cebulski on why he chose to use the pen name Akira Yoshida in the early 2000s to write a bunch of "Japanese-y" books for them.'[7]

Cebulski, Yoshida and Marvel were now in big trouble, and they knew it. The environment had switched from uncomfortably warm to blisteringly hot – it was time, they realized, for the truth. Cebulski *was* Yoshida; Yoshida *was* Cebulski. Mike Marts, incidentally, had got his wires hopelessly crossed: he had met and misidentified a Japanese translator. More senior figures, who *were* in the know, had hushed the scandal up.

The extreme heat of controversy and indignation had laid bare an entirely new reality. Yoshida and Cebulski had, for years, been known as two completely different people. Now, in these fiery surroundings, all looked different: there was only one person there after all. Yoshida and Cebulski – as unlikely as it might once have seemed – had been well and truly *unified*.

Just how far can we take this?

The bizarre story of Cebulski–Yoshida is weirdly useful in helping us to understand where unification ideas in physics went next – for it was temperature that turned out to be key. Initially, just about everyone had hoped that the infinity-busting technique of renormalization would work for the other three forces in the same way it had with electromagnetism. The excitement was palpable, for if it did, they would be done – as in go-home-and-put-your-feet-up-forever done.

The weak nuclear force, however, shattered this optimism – it simply refused to play along. Its infinities would not go away quietly, even under renormalization; for quite a while, the community was stumped. Then, in the mid-1960s, a bold new insight emerged.

Independently, Steven Weinberg and Abdus Salam (1926–1996) wondered if combining the key ideas of the electromagnetic and weak forces to form a whole new set of equations might resolve the dilemma. They strongly suspected any such hybridized formulae *could* be renormalized. In his book, *Dreams of a Final Theory*,

Weinberg admits: 'Both Salam and I had stated our opinion that this theory would eliminate the problem of infinities in the weak forces. But we were not clever enough to prove this.'[8]

He is being hard on himself, of course. It is true that others weighed in and refined their concept, but there is little doubt that Weinberg and Salam should take most of the credit. Thankfully, their novel idea came complete with its own testable consequence: it predicted the existence of three more particles. Each of these was a boson, the previously mentioned force-carriers, and each was later discovered. As a result, the duo won the 1979 Nobel Prize (along with Sheldon Glashow) for their work. Weinberg sums it all up like this:

> [Our idea] turned out to be a theory not only of the weak forces, based on an analogy with electromagnetism; it turned out to be a *unified theory of the weak and electromagnetic forces* that showed they were both just different aspects of what subsequently became called an *electroweak* force.[9]

A century before this, electricity and magnetism had been bound together by Faraday and Maxwell. Now Weinberg, Salam and Glashow had repeated their trick; electromagnetism was twinned with the weak force.

In fact, their theory made an even deeper claim: if the surrounding temperature got high enough, the two forces would become indistinguishable. In a superhot cosmos, electromagnetism *was* the weak force; the weak force *was* electromagnetism. The previously clear distinction – think Yoshida and Cebulski – literally melted away. What remained was *one force*, not two.

The beauty of this discovery is that universal temperatures really do get that high – provided we go back far enough in time. The Big Bang theory (see Chapter 4) says the universe was once much, much smaller than it is now – but contained the same amount of energy.

This made it unimaginably hot; there actually *was* a time when the electroweak force existed.

Later, as the universe cooled, the situation changed. The electroweak force split into two, each fragment having its own type of physical behaviour. One shard became electromagnetism and the other the weak nuclear force. Before this early cosmos separation, however, there were not *four* different forces in nature – there were *three*.

The electroweak unification sparked a frenzy of excitement among Big Thinkers, for obvious reasons. What if we go back *further* in time, they asked, to when the universe was even smaller and even hotter? Is there another great merger to be found? Will three forces turn into two?

Dare we dream, even, of pushing this argument to its dramatic conclusion and hope that there is a point where we hit only *one force* – a standalone equation capable of describing everything there is? Hawking, mulling over these questions, once again wears his heart on his sleeve: 'People have therefore sought a theory of everything that will unify the four [forces] into a single law that is compatible with quantum theory. This would be the holy grail of physics.'[10]

The grail-hunt has certainly gone well at times. The strong force was quantized in the 1950s, giving it a successful quantum version – one that came with those all-important force-carrying bosons. This theory is called *quantum chromodynamics*, or QCD, and its familiar form gives much hope to Big Thinkers. Many theoreticians are convinced this quantized strong force will fuse with the electroweak force at high enough temperatures, providing what is known as a *grand unified theory* (GUT) – meaning only gravity would remain in the cold.

Conviction is not always enough, however. It has been over five decades since Weinberg and Salam made their breakthrough, but still no GUT has proved valid. The best candidates predict that protons (particles found in all atoms) will, on occasion, decay into other particles. And, although this has been looked for again and again, it has, rather depressingly, never been seen. All the evidence suggests that protons just stay as protons. Sadly, turning four forces into three (unless we count Faraday too and say five into three) is as far as anyone has got.

Since the mid-1970s, then, scientists have opted for a compromise known as the 'standard model' (see Plate 10). It places the three quantized forces (electromagnetism, the strong nuclear force and the weak nuclear force) into a neat table, listing their respective particles' masses and charges, as well as their bosons. This visually tidy layout allows for continued belief in a profound link between them all – even if we don't know what it is.

Gravity, however, has not yet been quantized – so it doesn't fit into the standard model table. This inconvenient fact doesn't stop some textbooks from trying to shoehorn it in; they position it awkwardly off to one side with conspicuously empty boxes, as if it was an unwelcome relative in a grin-and-bear-it photoshoot. This clearly makes Hawking rather miserable:

> Since earlier observation had also failed to support GUTs, most physicists adopted an ad hoc theory called the *standard model*, which comprises the unified theory of the electroweak forces and QCD as a theory of the strong forces. But in the standard model, the electroweak and strong forces still act separately and are not truly unified. The standard model is very successful and agrees with all current observational evidence, but it is ultimately unsatisfactory because, apart from not unifying the electroweak and strong forces, it does not include gravity.[11]

After a more-than-promising start, unification seems to have run out of steam. At the end of the last chapter, however, we promised some *hope* – and from Hawking himself, no less. What was the basis of that promise? What did the great man do to keep the door open on a theory of everything?

The answer is really quite an odd one. Hawking caught his glimpse of salvation when he was looking at one of the most destructive forces in the universe – our old friend, the black hole.

A new hope

Hawking's dissatisfaction with the standard model is not his alone – the physics community as a whole has a love–hate relationship with it. On the plus side, there is something attractive about its compact tabular form; the three forces sitting next to each other give the impression of being a happy little family, with the electroweak unification hinting at yet deeper underlying relationships. The negatives, however, are undeniable: much of the standard model is a mess.

First, it does not (and cannot) give any *reason* for the various masses and charges of the particles within it: there are no patterns; they must be determined by experiment; they must be entered manually. Second, the strong force has not been united with the other two, even if some insist it will be one day. Third, because the standard model is essentially quantum mechanics – it contains only quantized forces – there is simply no room for general relativity. That means, of course, *no gravity.*

Many have tried – and failed – to resolve these issues. Some believe, very deeply, that gravity must also have bosons to carry it around, just as the other forces do – they have dubbed these bosons *gravitons.* No one has found gravitons in nature and it's hard to build a complete case for them even in theory, but this does not mean that they have yet been abandoned: the overwhelming demand for a theory of everything keeps even the longest-shot concepts on the table.

The fundamental problem is, as we have already learned, that quantum gravity is tough going – it keeps throwing up impossibilities. Try as we might, it seems that any attempt to get QM to play nicely with GR leads to fallouts: we get nothing but space–time contradictions, infinities and singularities.

In 1975, however, Hawking did something remarkable: he calculated a result that no one (himself included) had ever expected to find. His work was ingenious, daring, elegant and controversial to the point of being instantly dismissible – but it included an unmistakable hallmark from nineteenth-century physics that no scientist worth their salt could ignore.

Hawking, somehow, had done it again: he had uncovered an idea that would change the way physicists would think about the universe for ever. What is even more astonishing, though, is the *way* he did it – *he combined QM with GR.*

Black holes ain't so black

The story of this incredible manoeuvre actually starts years earlier, in 1970. Hawking begins it in typically down-to-earth style in *A Brief History of Time*:

> One evening in November that year, shortly after the birth of my daughter, Lucy, I started to think about black holes as I was getting into bed. My disability makes this a rather slow process, so I had plenty of time.[12]

Hawking was wondering how the surface (or *event horizon*) of a black hole could be precisely mathematically defined. He decided that it must be the series of points where light was neither escaping it nor falling into it, but was, in his own words, 'hovering forever just on the edge'. Armed with this new definition, he could analyse the surface area of any black hole. He could also prove this value would *never decrease*; black holes could stay the same size or get bigger, yes – but they could certainly not ever shrink.

This was an interesting finding. There is another quantity in physics that behaves like this – *entropy*. Entropy has been known about for at least 150 years and is associated with both disorder and temperature. Its value within a system can only ever stay the same or go up – it can never go down. Following Hawking's result, Jacob Bekenstein (1947–2015) of Princeton took the audacious step of suggesting that the surface area of a black hole actually *was* its entropy. Hawking was not amused. In fact, he was really quite angry.

The blatant problem with Bekenstein's claim, he complained, was that it missed the point of black holes entirely. In his own words:

If a black hole has entropy, then it ought also to have a temperature . . . So black holes ought to emit [heat as] radiation. But by their very definition, black holes are objects that aren't supposed to emit anything.'[13]

Hawking soon expressed his formal disapproval by publishing 'The four laws of black hole mechanics' with Jim Bardeen and Brandon Carter.[14] The triumvirate were having none of Bekenstein's nonsense: black holes were unapologetically black.

Looking back on this whole situation later, though, Hawking shows refreshing honesty as he describes where matters went next:

I must admit that in writing this paper I was motivated partly by irritation with Bekenstein who, I felt, had misused my discovery of the increase of the area of the event horizon. However, it turned out in the end that he was basically correct, though in a manner he had certainly not expected.[15]

Perhaps softening to the idea that black holes might actually permit some sort of escape, Hawking carried on chewing it over. Inspired by a conversation with some colleagues in Moscow, he sat down to do some more maths.

This time, he considered the core principles of both QM and GR – two theories that are supposed to be wholly incompatible – and came to an astonishing conclusion: maybe, he declared, 'Black Holes Ain't So Black' after all.[16]

Hawking radiation

We have said before that the uncertainty principle sits right at the heart of quantum mechanics. In a similar way, black holes are central to the equations of general relativity – they are an inevitable consequence of Einstein's equations. Since QM and GR are opposed in their very essences, one might think mixing the uncertainty principle with black holes would yield infinitely incomprehensible nonsense. The irrepressible Hawking, however, tried it anyway – and made it *work*.

The paper that emerged from his thinking appeared in 1975 in *Communications in Mathematical Physics*. Entitled 'Particle creation by black holes', it may well be Hawking's single most important contribution to physics. Indications that he is about to rewrite the rulebook entirely don't take long to turn up – the abstract, right from the beginning, is dynamite: 'In the *classical* theory black holes can only absorb and not emit particles. However it is shown that *quantum mechanical effects* cause black holes to create and emit particles *as if they were hot bodies with temperature*.'[17]

This is jaw-dropping. Hawking is going to show that black holes can release particles, something previously deemed impossible. What's more, he is going to tame these lions of GR using the chairs and whips of QM. Finally, he is going to undergird his case with a result that has been well understood for 200 years: the pattern of black hole particle emission will be more than just vaguely familiar – it will be identical to that of a *kitchen oven*.

Unsurprisingly, the paper is highly technical, but the basic idea of what Hawking did is not too difficult for even the layperson to understand. This is because an alternative picture of the same physics has arisen in the time since his original breakthrough – one that is easier to grasp and equally as valid. Hawking himself now uses this second approach in his popular work; we will do the same. Amazingly, we shall only need two ingredients: QM's uncertainty principle and GR's black hole.

Between them, these two provide everything we require to get our heads around Hawking's discovery. From the black hole we get its *event horizon* – the surface from beyond which there can be no escape, not even for light. From the uncertainty principle we get *virtual particles* – a writhing sea of particle–antiparticle pairs fizzing in and out of existence from 'nothing'. To appear, they 'borrow' energy from space – then pay it back again with their almost-immediate demise.

Hawking realized that the energy these mayfly-like particle-pairs required could be sourced from the black hole itself. The space around its edge would therefore be teeming with them, each duo an energetic mini-loan from the black hole's rich gravitational resources. The loan would be returned in full in a near instant, of course, so the black hole appears – at first glance, anyway – to be completely unaffected by this sideshow. But it *isn't*.

Every now and then something astonishing would occur: a particle–antiparticle creation event would happen right on the event horizon itself. One of the virtual partners would cross the point of no return and be sucked, for ever, into the black hole; the other, heading in the opposite direction, would get away (see Plate 11).

This bereaved survivor, according to theory, *cannot* disappear again – the vanishing act is a two-particle job and requires the presence of its teammate. Instead, then, it would begin a long, lonely journey across the universe – taking a tiny amount of the black hole's energy with it.

From a physics point of view, this is the exact equivalent of the black hole having emitted that solitary particle from inside itself – it was as if it had *let it escape*. Even Hawking did not believe this could be true; he hoped to find an error or loophole somewhere by which he could prove himself wrong. As he recounts in *Brief History*:

> When I did the calculation, I found, to my surprise and annoyance, that even non-rotating black holes should apparently create and emit particles at a steady rate. At first I thought that this emission indicated that one of the approximations I had used was not valid. I was afraid that if Bekenstein found out about it, he would use it as a further argument to support his ideas about the entropy of black holes, which I still did not like.[18]

What eventually persuaded Hawking and others, though, was what this steadily emitted energy (now called *Hawking radiation*) would *look like*. It formed a curve that was all-too-recognizable: black holes, it would seem, radiated energy in precisely the same way that any other hot body does – like ovens or kettles or lamps. His theory therefore felt 'safer' somehow: it had received a reassuring stamp of approval from the long-established and successful field of thermodynamics. To his initial amazement, Hawking could even calculate a black hole's *temperature* – that in turn meant that they would have entropy. Bekenstein had been right all along.

Securing a legacy

We have said that this might be Hawking's greatest one-off achievement – but *why*? So what if black holes aren't so black? Why does Hawking radiation matter? Why are physicists everywhere still in awe of this paper? There are at least three good answers to these questions and each one, on its own, would be significant. Together, they secure quite the legacy.

First, Hawking radiation demands a complete rethink about what black holes actually *are* – no longer are they exclusively one-way, their contents stolen away from the rest of the universe for all eternity. There is, Hawking showed, a *way out* – a very weird, unexpected and unconventional one, yes – but a way 'out' it remains nonetheless.

Second, it gave rise to another huge controversy in physics – one that added to the ever-increasing sense among theoreticians that they were missing something both deep and important in nature. This new and thorny issue was dubbed the *black hole information paradox*; we shall touch on it again very shortly.

Third, though – and probably most importantly – it is a meaningful real-world result that is widely accepted to be correct, yet it draws deeply on the fundamentals of both general relativity and quantum mechanics. Hawking had found a way of including both sets of rules while also ending up with a sensible, calculable answer that was consistent with reliable, age-old physics.

In other words, he had – in this strange and highly specific instance – *united* QM with GR. He had lifted the lid, just a little, on quantum gravity – allowing his colleagues the chance to sneak a peek of their own. And, even for that tiny opportunity in isolation, they shall all be eternally grateful.

Trouble in Paradise

Although Hawking had managed to bring the two warring titans a little closer, he had not resolved the dilemma completely. To be brutally honest, he had only really scratched the surface; the problems and issues of infinities and backgrounds (see Chapter 5) were still there.

Soon, as the physics community wrapped their brains around what Hawking radiation actually meant, they began to realize that it came with a highly disturbing corollary. Rather than improving QM–GR relations and perhaps even ushering in a new age of peace, the Cambridge cosmologist might actually have made matters *worse*.

In theoretical physics, the word *information* has a very specific meaning. It refers to knowledge about the universe and it allows us to make predictions about the future. One of the central tenets of quantum mechanics is that the *total amount of information in our cosmos remains the same* – information can never be lost, even if it does change its format at times.

A common analogy used to explain this idea is that of a burning book. Before it is set alight, it hosts reams of information – afterwards, as a pile of ash, it (apparently, at least) holds none. Surely, one would conclude, that information has been lost in the fire. 'Not so!' is the shout of the physicist.

Every printed page of the book, this genius explains, can be reconstructed: all we need do is track the particles. Some now sit in the grate while others drifted off as smoke – but all of them can still be found *somewhere*. And crucially, when we find them, we can analyse them.

By studying their current states, we can find out about their immediately preceding ones – for the laws of physics are time-reversible. If we carefully apply this principle to each and every particle, we can painstakingly rebuild the book using maths – and then, of course, we can read it. According to this argument, no information is ever lost; it is stored in the histories of each tiny grain of the universe.

This may seem highly improbable, but it does turn out to be correct in every scenario that scientists have tested it in. Admittedly, they cannot put burnt books back together, but they *can* devise small experiments that explore the same ideas. The principle always holds true – the belief that information is conserved for ever has become one of the cornerstones of all modern physics.

The equations of QM, for example, work on that basis – as do others. If it turns out that information *isn't* eternally protected, physicists would be in quite the pickle: they would face the very real prospect of their subject being torn top from bottom. Information loss is therefore bordering on the unthinkable – but no one, it appeared, had told black holes.

Hawking (in the very same paper) was able to prove that a certain class of black holes was doomed. For these poor gravitational souls, the eventual consequence of losing energy as particles would be their complete evaporation – they would, after radiating for a sufficiently long time, simply cease to exist. The total disappearance of a black hole, however, is more than just mildly melodramatic – it is, in fact, a scientist's nightmare.

It had long been known that black holes could take in information – it would simply be pulled in just like anything else. That information (from our book, say, if that fell in) would then be hidden from outside view. Astrophysicists had thought about this and decided that they were OK with it: the information still clearly *existed* – the issue was only a matter of *access*. Hawking's portentous assertions, though, tore away this cosmological comfort blanket – as he himself was all too aware: 'If, as I showed . . . quantum mechanics allows black holes to lose their mass and disappear, there is a difficulty. After the black hole is gone, what has then happened to the information?'[19]

This more-than-a-little-perturbing question remains open even today – for no one knows where the information has gone. Lee Smolin (one such worrier) sees the enigma as one of the (many) ways in which the universe is telling us we need a whole new theory altogether:

The quantum description of the world is supposed to be exact, and there is a result implying that when all the details are taken into account, no information can be lost. Hawking made a strong argument that a black hole that evaporates away loses information. This appears to contradict quantum theory, so he called this argument the black-hole information paradox. Any putative theory of quantum gravity needs to resolve it.[20]

Once again, then, we find ourselves in wait of quantum gravity. The search for what Hawking calls the 'holy grail of physics' started off pretty well – QED, the electroweak theory, QCD – but has run into dead ends ever since. The taped-together compromise of the standard model hangs over the discipline like an albatross and even Hawking's magnificent QM–GR project of black hole radiation hasn't helped for, despite its successful use of both world views, it also threatens to break all of physics.

Many, at this point, have surrendered – the cosmos just seems one step ahead. For Hawking, however, surrender was not an option; he would have yet another stab at the prize. He was done, though, with chasing down small fry – even the mighty black hole was now deemed too puny. This time, he would think without limits: Hawking was about to go *big*.

6

Part 2

Mathematical dogfights • Imagine that • Waves, particles, probabilities, histories • Hawking goes big • A disclaimer • Extraordinary measures • Extraordinary times • A theory of (almost) everything • Multiverses, metaphysics and myths

Mathematical dogfights

A few centuries ago, professional mathematicians could be quite the talk of the town. Occupying an odd social niche that sat somewhere between superstar celebrity and gun for hire, these number-crunchers were often in great demand, enjoying an elevated (and well-paid) status in society – especially in the powerful provincial courts of mid-Renaissance Europe.

With their exalted role, as one might suspect, came great risk: mathematicians were expected to justify their standing to their rich patrons by proving they were the best of the best. On occasion, a rival might take advantage of this and challenge an incumbent to a 'duel': each would set the other problems to determine who was the real top dog. The victors, of course, gained yet more fame and fortune; the losers limped off to lick their wounds. Neither, however, could ever truly rest – for there were always more tricks to be learned.

Many of the puzzles posed in these dogfights were carefully structured algebraic conundrums; they were cracked by using well-known but long-winded techniques that went back hundreds or even thousands of years. The real skill, therefore, was in choosing and then following the appropriate method through to its end without making a single mistake along the way. The more methods one was aware of – and the more careful one was in using them – the better.

Every now and then, though, this status quo was disturbed – some gifted (or lucky) individual would stumble upon a diamond in the mathematical rough, unearthing something genuinely revolutionary. With such novelty came significant advantage: new and nifty shortcuts could allow previously essential stages to be skipped; new and nifty riddles (especially ones with unusual or unexpected results) meant duellists could throw their opponents hopelessly off the scent.

Of all the various breakthroughs that a calculator might make in their own time, however, one was far greater than any other: the discovery of an answer to a problem that everyone else had thought was unanswerable. These immediately top-secret solutions were the gift that kept on giving: so much so, in fact, that some old guns even went to the grave without ever passing them on. Find one, therefore, and everything changes – that is precisely what happened to Rafael Bombelli.

Imagine that

A no-nonsense engineer from Bologna, Bombelli (1526–1572) had been studying *cubics* – think quadratics, but with an extra level of multiplication. A certain set of these cubics had already been written off as 'irreducible': the normal technique could get a puzzler so far, but no further. This was because, at a specific point, it asked for *the square root of a negative number* – an idea that was patently absurd.

To see why, let us consider the options. We are looking for a number that, *if we multiply it by itself,* gives a *negative* result. This number, therefore, cannot be positive or negative, for both will give positive answers. The only remaining candidate is zero – but zero times zero is *zero*.

At this juncture the mathematicians gave up, because the truth is both obvious and unavoidable: such a number *does not exist*. These particular cubics, it was clear, were a lost cause – after all, how could a number that was not negative or positive or zero even be conceived of, let alone inhabit reality?

Bombelli, the engineer, thought differently. He felt that, even if the signal-box along the way was broken, the train could still make it to the station. As such, he took a truly remarkable step – he *pretended* that these non-existent numbers *did* exist and carried on with the method regardless. To his astonishment and delight, he found that he really could get all the way to the answer – but how could this possibly be?

It turned out his pretence had paid off: the non-existent numbers he inserted in the middle of the process made an exit again (by cancelling themselves out) before the end of it, leaving a real and sensible answer. Sure, he had gibberish on his page partway through, but everything resolved itself in time for the solution to make sense anyway. For a while, Bombelli struggled with the validity of what he had done and wondered if it could really be considered to be maths. Was he – somehow – cheating?

Eventually, however, he came to view these new and mysterious entities – now known as *imaginary numbers* – as powerful tools for solving previously intractable problems. Despite the huge scepticism of some of his colleagues, Bombelli decided to make his peace with the radical approach. In 1572, looking back on all this, he wrote: 'It was a wild thought in the judgement of many; and I too for a long time was of the same opinion. The whole matter seemed to rest on sophistry rather than truth.'[1]

In the time since Bombelli penned this in *L'Algebra*, his imaginary numbers have become a vital and established part of everyday mathematical practice. They appear in multiple areas of the discipline and even make their way into the tangible world of physics and engineering. Analysis of electric circuits, for instance, often demands the use of an imaginary figure or two – but they always drop out again after a bit, ensuring that a final value for voltage or current is as real as real can be.

Quite what these imaginary numbers actually *are* (and why they appear in real-world physics) is still a matter of debate. What is clear, however, is that Bombelli and his dogged persistence changed maths and science for ever. His reward for such an achievement? Well, he has a moon crater named after him, of course.

Oh, and Stephen Hawking decided to use his work to find a theory of everything. There's that too.

Waves, particles, probabilities, histories

The stranger-than-fiction link between Bombelli and Hawking is rooted in the madness of quantum mechanics. The reader will recall that one of the key steps in the development of QM was the realization that, in the world of the small, classical physics is blurred: everything can be thought of as a-little-bit-wave-and-a-little-bit-particle.

A whole new science emerged from this indistinct mess: physicists realized that they could analyse any tiny, self-contained system by writing down a single mathematical formula – one that they called the *wave function*. As the name suggests, this equation is wavelike in form and it holds within it all the data about the set-up – including *every possible state* that the system could ever be found in.

To make a prediction about how that system will behave, theoreticians perform a particular calculation on the wave function: they ask it for the *probability* of any particular result. If there is no chance whatsoever of that outcome arising, the answer they get will be *zero*; if it is a nailed-on certainty, they will get *one*.

For everything else – all the it-might-happen-it-might-not results – the value will be somewhere in between zero and one, with bigger meaning more likely. As a reminder of how this works, we can consider once again the lone electron of Chapter 3 – the one hurled towards a barrier with two tiny slits in it.

In the electron double-slit experiment, the key piece of information we are interested in is this: where will the electron eventually strike the screen? It is emitted from an electron-gun (let's call this A) and travels through the slits to a single point on the screen (B). The difficulty of all this is that the electron, as it approaches the slits, seems to change its behaviour altogether: it stops acting like a particle and switches instead to wave mode.

This is why the wave function concept is so useful. For this experiment, the wave function need only contain the bare numerical facts about the electron, slits and screen. It is entirely unfussed by the seemingly indeterminate nature of the electron – everything is taken care of by the maths.

Once the wave function is built (i.e. given its data), we are ready – we can ask it where the electron will finish up. Essentially, we pick a location for B and run the calculation; we will get somewhere in the range of 0 to 1 for the answer. Note the lack of a classical guarantee – the wave function only gives a probability. It does the same, of course, for any other value of B we might be interested in: which means that wave functions give an infinite number of possible answers, each of which is probabilistic.

Infinities and probabilities may seem wholly unmanageable, but it isn't quite as bad as it seems; these predictions are testable by experiment. If, for example, the wave function states that the electron will hit a specified point B one time in every thousand, we can check this. How? By running the exact same experiment *lots of times* – and finding out if this probability holds up. Sure enough, it does – wave functions are always spot on.

Herein lies both the beauty and the success of QM – it deals with minuscule and simple situations; it gives answers as verifiable probabilities; its easily doable experiments can be repeated again and again; it is proved right on every single occasion. And, at its core, sits the all-conquering wave function – a well-oiled prediction machine.

There is, however, a downside: the quantum mechanical maths of even slightly larger systems than the electron double-slit can quickly get out of hand. Calculations explode to become prohibitive, leaving the professionals in desperate need of (rarely available) shortcuts.

One can see, then, why Richard Feynman's *sum over histories* method seemed such a Godsend when it arrived: for he had found a different way to think about the infinite number of answers contained in the wave function. Why not imagine, he said, that everything the wave function said *might* happen actually *had happened* – and then add all these everythings together? Hawking puts it like this:

> In this approach, the particle is not supposed to have a single history or path in space-time, as it would in a classical, nonquantum theory. Instead, it is supposed to go from A to B by every possible path ... The probability of going from A to B is found by adding up the waves for all the paths.[2]

Astonishingly, and despite the fact that it sounds overwhelmingly bamboozling, Feynman's fudge on the wave function *worked*. Nearly all his infinite alternative 'histories' went on to cancel out; as they did so, they left only the correct probability behind.

Doing the calculations this new way was far easier than using the original techniques; Feynman had simplified the process significantly. In turn, he had opened up the possibility of analysing wave functions for more complex systems. Thanks to him, QM – which had previously been confined to just a small class of pared-down scenarios – now had a (slightly) bigger reach.

Hawking goes big

Back in the mid-1960s, Hawking had announced himself on the cosmological scene by plonking down a singularity right at the start of the universe. Horrifyingly, this meant that the origin of our world was for ever shrouded in mystery; the infinities Hawking had placed there destroyed any chance whatsoever of analysing the first moment in time. We would, his PhD had trumpeted, never know where we had come from.

Despite this being his own doing, Hawking became desperate to overturn it. We shouldn't be all that surprised by this; he was a man after solutions and a singularity is pretty much the opposite. Could it be made to go away somehow?

Hawking strongly suspected that, if the cosmos could indeed be unlocked, quantum mechanics would turn out to be holding the key. He also believed that a theory of everything would involve quantized versions of all four forces of physics. Over time, then, he began to obsess over one definitive question – could the universe, *in its entirety*, be made to succumb to the equations of QM?

The obvious and overwhelming answer to this question – at the time that Hawking began asking it – was 'No'. QM had been built, very specifically, to deal with the small and the simple; the very notion of the wave function is that it keeps every detail of a system inside itself. Provided this is just one or two particles in complete isolation from the rest of the world, it remains doable. Push it much further, though, and all bets are off.

Take the flip of a single coin, for example, as a suitable cautionary tale: the coin contains ten thousand billion billion atoms, and not one of these can be called self-contained – each is subject to outside forces from our hand, from the air and from the gravity of the Earth. Writing a wave function down for a coin toss, therefore, is not just tricky – it is *impossible*.

What's more, even if we could write such an equation, we could never actually test it. The predictions of a wave function, as the reader will recall, are given as *probabilities* – to check the validity of these, the experiment must then be repeated millions of times. For our coin flip, this is once again hopelessly beyond us: every flip will be at least slightly different in one way or another (thumb speed, room temperature, air circulation, etc.) so it cannot truly count as a repeat. There are no two ways about it: working with QM, on any macroscopic scale, is too tough.

Hawking, however, was undeterred by such trifling technicalities as these. He was looking to eliminate singularities and uncover a theory of everything; making excuses and qualifications was not in his playbook. For him, unification was the name of the game and nothing less would suffice. Why worry about what people said couldn't be done? After all, he had already broken the so-called 'rules' once with his black hole paper – in which he had 'united' QM and GR.

Faced with this new challenge of adapting QM to the universe – or perhaps the universe to QM – the ever-optimistic cosmologist decided to look back on his previous work for much-needed encouragement. Mulling over the fate of black holes, he realized he might be able to harness the consequences of Hawking radiation for a new cause – for he saw, deep within his theory, a clue as to what to do next:

> The most likely outcome seems to be that [a radiating] black hole will just disappear . . . taking with it . . . any singularity there might be inside it . . . This was the first indication that *quantum mechanics might remove the singularities* that were predicted by general relativity.[3]

This is all very well – but how could QM ever be made to work for anything bigger than a handful of particles? If even something as mundane as a coin toss was beyond the pale, what was the point of aiming any higher than that?

Hawking desperately needed a new trick – something to grab hold of the out-of-reach calculations and drag them down into the realm of the scientifically achievable. One day, however, it hit him: the solution was not a *new* trick after all – instead, it was an *old* one:

The methods that I and other people were using in 1974 were not able to answer questions such as whether singularities would occur in quantum gravity. From 1975 I therefore started to develop a more powerful approach to quantum gravity *based on Richard Feynman's idea of a sum over histories.*[4]

Hawking was beginning to put it all together. He had long known what he wanted to do: unify QM and GR; get rid of the singularity at the beginning of time; formulate a theory of everything. Now, he knew how to do it: he would apply Feynman's sum over histories methodology to the cosmos.

After years of battling back-and-forth with the physics, he finally got there. Working and writing with James Hartle, Professor of Physics at the University of California, Santa Barbara, Hawking published his answer. The title of their 1983 paper? Well, it was no less than this: 'Wave function of the universe'.

A disclaimer

Hawking, characteristically, decided to go for it right from the start – he promised a revolution from as early as line one: 'The quantum state of a . . . universe can be described by a wave function . . . we put forward a proposal for the wave function.'[5] There really is no way around this: the full mathematical construction of the Hartle– Hawking (HH) proposal is more than a little daunting – in fact, in it is fiendishly complicated. Today, more than three decades later, highly trained specialists still argue about it; not simply about whether it is correct or not, but even about what it means in the first place. The quantum-cosmological jury is still out; it has befuddled the brightest and best.

It is, therefore, somewhat tempting – in a book like this, at the very least – to give the HH universe a super-wide berth; to smile, nod and move on. There are two good reasons, however, for not doing that – and they are more than sufficient for us to press on. The first is that the basic idea is actually fairly accessible, even if the scientific detail then gets tough. The second is that Hawking chose to make the proposal the centrepiece of his two biggest works – both *Brief History* and *Grand Design* are built around it.

So, then – and with the disclaimer put out there that some of this will be stretching – we shall hunker ourselves down nevertheless; and we shall give it a really good go.

Extraordinary measures

In taking on a task that anyone and everyone had already deemed unachievable – or would have done had they ever actually entertained the prospect, which they almost certainly hadn't – Hawking and Hartle knew they would have to make some fairly major concessions.

Wave functions, after all, were designed to be used on tiny, stripped-down, non-complex systems, that means they are hardly appropriate for a *universe* – especially not a universe as rich and varied and complicated as ours. There is simply no way to include all its data – all its possible states – and Hartle and Hawking knew it.

Their first compromise, then, was to radically and drastically streamline what they actually meant when they referred to the 'universe'. In the paper, Hawking says:

> After a general discussion of this proposal for the ground-state wave function we shall implement it in a *minisuperspace model*. The geometrical degrees of freedom in the model are restricted to spatially homogenous, isotropic, closed universes with S^3 topology.[6]

There is no need to panic – this is not quite as horrendous as it sounds. Essentially, Hartle and Hawking are making everything much, much simpler to work with. *Homogenous* and *isotropic* (as we saw in Chapter 4) basically mean the universe looks the same in every direction from anywhere, and the *closed S^3 topology* part limits their universe to a sphere of finite size.

This *minisuperspace model*, therefore, has had nearly all its features removed – it is a bland, unremarkable husk that will act as an all-too-necessary substitute for our far-too-plentiful-and-far-too-real actual existence. Sure, the two men would have preferred to calculate values for the genuine article instead – but we just can't fit it into QM.

If Hartle and Hawking's first step was to simplify the *model*, their second was to simplify the *method*. As previously mentioned, they would evaluate their wave function using Feynman's sum over histories – it made everything that little bit easier. To make this work, however, they needed to change the maths. It was a change that was seriously weird – and one that called on a ghost from the past.

Extraordinary times

When Einstein sketched out his relativistic universe, one the main features was a new entity altogether: *space–time*. His equations bound space and time together so that the two were now somehow inseparable. Crucially, however, they were not quite the *same* – for his maths involved treating the three spatial dimensions differently from the one of time, even if the difference was quite subtle.

This way of handling things can be written, in mathematical shorthand, as [+, +, +, -]. Here, the three pluses are dimensions of space, the lone minus the dimension of time. The sudden switch in sign should serve as a constant warning to overly enthusiastic physicists: there may well be something called space–time, yes – but don't go tarring space and time with the same cosmic brush. The point is an important one and mathematician Paul Nahin presses it home quite firmly: '[The minus sign on time] also makes it pretty clear that time really is fundamentally different from space, a point that many science writers today have muddied in overly simplistic popularizations on relativity theory.'[7]

It is here, though, that Hawking and Hartle found a way to change the game. Realizing that they were very likely to get stuck in their attempt to use the sum over histories method for their universal wave function, they opted for a crazy work-around – one that had been four hundred years in the making. Channelling their inner Bombellis – and perhaps, like him, resting 'on sophistry rather than truth' – they would have to do some pretending of their own.

Hawking drops this bombshell in surprisingly matter-of-fact manner, recounting in *A Brief History*: 'To avoid the technical difficulties with Feynman's sum over histories, one must use *imaginary time*. That is to say, for the purposes of the calculation, one must *measure time using imaginary numbers, rather than real ones*.'[8]

Boom! This is an extraordinary statement and worth dwelling on for a (real) moment: Hawking and Hartle were using a mathematical trick from the Renaissance to get themselves out of a near-impossible quantum-mechanical predicament. Quite what imaginary time actually *is* remains a separate question entirely, of course, but for now it was simply a tool. In short, it played the role of quantum-cosmological duct tape: it was undeniably a fudge, yes – but it just about held things together.

In fact, as it turned out, it did better than that. Switching to imaginary time had a second consequence and one that was near-miraculous in its significance for Hawking. Amazingly, Einstein's rules about the four dimensions of our universe switched too: they now read [+, +, +, +]. This was a new type of space–time – one in which time was no different from space.

Indeed, 'no different' does not go far enough: what has actually happened is that time, as a distinct and separate quality, has ceased to exist – it has been transmogrified into a fourth dimension of space. The behavioural and conceptual boundary between the two is no longer just blurred; it has *disappeared altogether*.

What did this stupefying space–time shift signify? What difference could this lack of difference make? The answer was truly astonishing – and it all came down to the *shape* of the cosmos.

A theory of (almost) everything

Armed with their metamorphosed space–time, Hartle and Hawking were nearly there. Rather than their slimmed-down minisuperspace model boasting three-space-and-one-time as its underlying dimensional structure, it effectively now had four-space scaffolding – it contained *no time* whatsoever.

A universe like this is known as *Euclidean* – because it is reminiscent of the high school geometry of Euclid. It is, as bizarre as it seems, genuinely timeless; all its dimensions behave like space. The result of this shift, from Hawking's perspective at least, was profoundly joyful: without time, there can be no beginning. Without a beginning, *there can be no singularity*:

> In the classical theory of gravity, which is based on real space–time, there are only two possible ways the universe can behave: either it has existed for an infinite time or it had a beginning at a singularity at some finite point in the past. In the quantum theory of gravity, on the other hand, a third possibility arises. Because one is using Euclidean space–times . . . it is possible for space–time to be *finite in extent and yet have no singularities* that formed a boundary.[9]

Hawking and Hartle, in seeking to run Feynman's physics across their universal wave function, had managed to ditch the nasty infinities at the beginning of time. This was because they had done a Bombelli; they had defied the conventional wisdom.

Their universe – unlike all predecessors – 'began' in imaginary time; the result was a four-space with no hint of a singularity (see

Plate 12). It also incorporated the core ideas of QM (it had a wave function) and the tenets of GR (it had space–time). Had the pair, then, finally cracked it? Was this – whisper it quietly – a theory of everything?

Not quite. The first issue at hand is to ask how we get back to a meaningful universe. After all, our clocks don't measure imaginary time and we don't live in Euclidean space. It is all very well using imaginary numbers as a fiddle, but it won't get us anywhere in reality.

This problem runs even deeper, in fact. In the HH universe – also sometimes called the *no-boundary proposal* – the singularity only really disappears in the imaginary time regime. View the same situation in real time and it reappears. This is undeniably problematic: is it there or is it not?

The fuzzy compromise from the no-boundary proposal is to start off with imaginary time to avoid the singularity, then – after an arbitrary and unspecified span – to switch to real time, allowing our universe to evolve normally as per the original Big Bang model. If this seems unsatisfactory, that's because it is. Such is the nature of cutting-edge, first-iteration science.

A second issue is whether or not an idea like this that (sort of) works for a super-basic, monochromatic minisuperspace model can ever truly be beefed up enough to deal with our beautiful and multi-faceted actual home. No one (currently, at least) has even the remotest idea how to do so; does that mean that it cannot be done?

Third, it is worth mentioning again that the sum over histories technique needs three inputs to do its stuff. It needs a starting point (A), a finishing point (B) and all the possible routes between the two. At this point, we can be slightly more specific: by 'possible routes', what we really mean is 'the rules of the system that dictate all the possible routes'. Essentially, then, these rules are the laws of physics.

For a given universe, the *A-rules-B* set-up is surprisingly simple to describe. *A* is how it began; *B* is its condition right now; the *rules* describe every permitted evolutionary path that leads from one to the other. Throw all this data into the wave function pot, stir it a bit, add everything up – and out will spring the probability of the existence, *in reality*, of B.

In their hugely ambitious project, however, Hawking and Hartle were stuck: they could only be sure of one of these three inputs. That was, of course, B – the present state of our universe – but even that, they could not actually use, because it was far too complex. Instead, they used their minisuperspace model. But what about A and the rules? What did they actually enter?

To answer this, we shall turn to an expert – one of only a handful of people in the world currently working on the quantum gravity that so fascinated Hawking. In point of fact, Aron Wall is based in Hawking's own Department of Applied Mathematics and Theoretical Physics at Cambridge and is therefore helping to continue his legacy. Discussing the no-boundary proposal and its use of Feynman's *A-rules-B* structure, Wall says:

> There *is no A*! A is replaced with *no space existing at all* – Hartle and Hawking do a sum over all four-dimensional Euclidean spaces with *just a single boundary: B*.[10]

This seems crazy. How can a universe not have a start?! How are we supposed to know what it was like at the 'beginning'? Well, this is precisely what is distinctive about the HH universe – since it is Euclidean and timeless, it doesn't really have a beginning, in the traditional sense, and the duo can therefore jump straight to the *rules* part of the process. In their system, the apparently missing A is sort of built into the rules directly rather than being its own distinct piece of information.

When Hawking and Hartle roll A up inside their rules, it is not as far-fetched as it seems. Indeed, it is not entirely dissimilar to the way that electromagnetism merges electricity and magnetism – or even how the electroweak interaction blurs electromagnetism with the weak nuclear force. Wall again:

In all previous models of physics, physicists had to specify *both* the laws of physics [*rules*], *and* the initial conditions [*A*]. The real achievement of Hartle and Hawking (if their proposal is correct) is that they *unified* these two topics, so that the laws of physics *also* determine the initial state (and vice versa).[11]

Still, this seems rather a stretch. So outlandish, though, is the notion of applying QM to the entire universe that the bar for what can be deemed a success is set very low indeed. Some may well read this and think the whole thing somewhat absurd – playing with imaginary time, arbitrarily switching back to real time, blurring away a beginning, trading an unrecognizable shell for our real cosmos – but what really matters is this: would Hartle and Hawking's ramshackle wave function actually hold together when somebody asked it for an actual *result*?

The answer, unbelievably, was 'Yes'. It didn't blow up to infinity; it didn't collapse into nonsense. It gave a non-zero probability for the existence of the (heavily modified) cosmos. It even hinted at a link to genuine reality: some of its outputs were suggestive of the cosmic microwave background radiation – a phenomenon that has been detected by real people with real antennae in the real world. Perhaps, against all the odds, Hawking and Hartle had *done it*.

On the other hand, though, perhaps they had *not*.

Multiverses, metaphysics and myths

The HH universe, it would seem, comes in literally countless versions. There are an infinite number of four-dimensional Euclidean paths that it can take, and all show up in the maths with varying degrees of probability. In short, the wave function of Hartle and Hawking does not so much predict the existence of *one* universe – it predicts the existence of *many*.

Hawking, naturally, is aware of this: he is more than open about the situation in *The Grand Design*: 'In this view, the universe appeared spontaneously, starting off in every possible way. Most of these correspond to other universes. While some of these universes are similar to ours, most are very different.'[12]

What on earth(s) are we to do with this statement? If a scientific model like this can predict *any* conceivable outcome, how can we ever decide if it is right? We are in danger, suddenly, of heading into contradiction: for in this case, the *same theory* (and the same results) can legitimately be interpreted as *correct* or hopelessly *wrong*. Let us take a brief moment to see why.

It is possible, we could argue, that Hawking and Hartle have nailed it. They have, after all, predicted the existence of a pseudo-universe that is, in some ways, similar to ours – all the other predictions in their work can therefore safely be ignored, even if there are an infinite number of them.

Alternatively, however, we could also make the case that their project doesn't work at all – the only reason it includes a prediction of *our* universe is because, by default, it includes a prediction of *everything*. Of course, then, it will predict ours, by default – their 'finding' is nothing but a truism.

Deliberations like this, though, are of a new kind altogether. They are no longer discussions *within* science, but instead are discussions *about* it – they have crossed the great divide, migrating from physics over to metaphysics. One minute, we were evaluating wave functions; the next we were pondering philosophy.

It is not all that surprising, then, that the no-boundary proposal is the one area of Hawking's work that drives him most towards his Big Questions. Time and again he links the HH universe to other transcendental ideas. This wide-ranging passage, drawn from *The Grand Design*, is a typical example:

Over the centuries many, including Aristotle, believed that the universe must have always existed in order to avoid the issue of how it was set up. Others believed the universe had a beginning, and used it as an argument for the existence of God. The realization that time behaves like space presents a new alternative. It removes the age-old objection to the universe having a beginning, but also means that the beginning of the universe was governed by the laws of science and doesn't need to be set in motion by some god. If the origin of the universe was a quantum event, it should be accurately described by the Feynman sum over histories.[13]

Whatever this is, it is certainly not astrophysics. Hawking, in just one paragraph, has mentioned ancient Greek thought, the nature of time, the existence of God, the nature of science, the first cause argument and even quantum mechanics. He is more than waist-deep in philosophy and theology here – a situation that occurs more often in his work than one might have previously thought.

Perhaps this is because unification – the grand project of joining seemingly disparate ideas together – seems to run in Hawking's very blood. Not content, then, with combining QM with GR in black holes (or even in universal wave functions), he wants to unify everything else there is too – myths, and God, and all.

And, although he may be out of his narrow field when doing so, Hawking is still undeniably a Big Thinker. His conclusions about all these matters – that involve the multiverse, Norse mythology and even goldfish – are both fascinating and hugely influential. In fact, they deserve a chapter of their own. With our next one, then, they shall get it.

7

Tyres and anchors • Counter-punches • Crossing the divide • Is philosophy dead? • You Kant be serious!? • Being critical about realism • A rose by any other name • One foot in each camp • Introducing the multiverse • Creating history • Hawking and God • Switching the spotlight

Tyres and anchors

In the sweltering heat of Kuala Lumpur's Merdeka Square any unnecessary physical exertion is not advisable. It is wiser – and especially so for European tourists – to find shade, sip at a glass of honeyed lime juice and watch the world go by. In the February of 2002, however, Mariusz Pudzianowski went for a different option: he chose to flip a tyre over four times and drag an anchor a few metres along the ground. He completed these two tasks in precisely 39.01 seconds – and, just like that, he launched a dynasty.

'Pudge', as he is affectionately known by his fans, was not in Malaysia on a journey of culturally catalysed semi-spiritual self-discovery – he had ventured there with a very different goal in mind. If all went according to plan, the squat and stocky traveller would be returning home with a brand-new identity: he would have become the World's Strongest Man.

Pudge was not alone in his quest. Nine other superhumans had joined him, most of whom were taller, heavier and considerably more experienced. At the tender age of 24, in fact, Pudzianowksi was a relative newcomer – but he was on the verge of becoming a legend.

The tyre weighed as much as a horse; the anchor was a third of a tonne. No one was able to shift them faster than the explosive youngster, who went on to win the next Herculean test as well. By the end of the tournament, the Pole was untouchable – he had redefined what the industry thought was possible. From his breakthrough in 2002 until his retirement in 2009, Pudzianowski was the ultimate Alpha Male – he ruled the giants' roost. He won Europe's Strongest Man on all six occasions he entered it and was the World's Strongest Man five times – an achievement that remains a record to this day.

At the age of 32, though, Pudzianowski had had enough – it was time for a change of career. After a decade of pushing his body to unimaginable extremes, one might guess the strongman would switch to something a little less strenuous – painting watercolours in the countryside, perhaps, or even a nice quiet office job. Once again, however, Pudge had different ideas – he was going to learn how to *fight*.

Entering into perhaps the only universe more saturated in testosterone than the one he had previously reigned over, Pudzianowski started training as a mixed martial artist. His first nearly-anything-goes bout arrived in December 2009, when he faced Marcin Najman, a former professional boxer. He won. In the first round. Via knockout. Najman held out just a little bit longer than the tyre and anchor: the fight (if it could really be called one) lasted 43 seconds.

Is this really all that surprising? Pudge had been, for quite some time, the strongest man on the planet – what better qualification could there be for climbing into a ring and beating the living daylights out of someone? Prior to his deconstruction of Najman, Pudzianowski had thrown out a bone-chilling reminder of his power to the rest of the roster: 'The left hand brings death – but the right one even I am afraid of.'[1] Surely, given his incredible pedigree, the writing was on the wall: Pudge would simply bulldoze a path to the top. Another World Title seemed a formality; after all, who could possibly hit anyone any harder?

Counter-punches

The website *Fight Matrix* is compiled by careful statisticians – they collect data on MMA contests from all over the world and rank the sport's fighters accordingly. Each weight category has its own extensive list of as many as 250 international pugilists; each fight is dutifully monitored and the results added into the database. Since destroying Najman a decade ago, Pudge has fought a further seventeen times – so is he now the Champ?

Err . . . no. In fact, he is languishing down at 139th position in the heavyweight rankings, has lost six fights and has failed to earn a match against anyone considered half-decent. Sure, he is making ends meet and he still enjoys a higher profile than most due to his previous exploits, but he is not exactly setting the MMA world alight. Which is strange. Because he is very, very *strong*.

The issue at hand is this: pure, unadulterated strength is indeed an asset in the ring – a major one – but it is not enough on its own. Pudzianowski's betters are those who have become specialists in the art; they have worked on, and perfected, a variety of brutal techniques and tricks that are simply unknown in the world of the strongman.

These grapplers are, therefore, battle-hardened: they can twist an opponent's arms or legs half-off; they can wriggle out of impossible situations; they can see a fist coming, dodge it and land a counter-punch before a regular mortal has even drawn breath. Pudge is undeniably *stronger* than his opponents, but this is not his game. He was utterly dominant in his own field, yes – but in theirs, he has often been found wanting.

Could there be a lesson here – in this story of a world-beating genius switching to a related field and finding it tough going – for Hawking?

Crossing the divide

When *A Brief History of Time* was released, the Cambridge-based heavyweight was at the height of his cosmological powers. By the time of its publication in 1988 his multiple displays of strength had stunned many. He had stuck a singularity at the beginning of the universe; he permitted escape from black holes; he had even written a quantum wave function for the entire cosmos. There can be little doubt about it – Professor Stephen Hawking was the Physics World's Strongest Man.

In that era, Hawking – publicly, at least – was still very much the scientist. He knew his work had the potential to cross over into other territories, but he himself held back from doing so in any heavy-handed way. One interview with American television is a good example of his restraint:

> It is difficult to discuss the beginning of the universe without mentioning the concept of God. My work on the origin of the universe is on the borderline between science and religion, but I try to stay on the scientific side of the border.[2]

This attitude – in which Hawking appears to be content to let philosophers deal with philosophy and theologians deal with theology – is also present, at times, in the text of *Brief History* itself. In its conclusion, he writes:

> Even if there is only one possible unified theory, it is just a set of rules and equations. What is it that breathes fire into the equations and makes a universe for them to describe? The usual approach of science of constructing a mathematical model cannot answer the questions of *why* there should be a universe for the model to describe.[3]

The thing about the God question, though, is that it is very hard to leave alone. The links to the idea of a Creator in Hawking's work are obvious and it would hardly be fair to deny him a say on the matter. He was clearly mulling it over; others were too. David Hickman, who produced a screen version of *Brief History* in 1992, insisted that this aspect of Hawking's physics was a significant part of its success:

> The most exciting thing about cosmology is the fact that it interfaces metaphysics and conventional science. It's very interesting that Stephen has attracted a lot of attention over the religious aspects of his work, as well as the fact that he is close to a number of physicists with deep theological concerns.[4]

In reality, it was inevitable that the paper walls dividing the physics, philosophy and theology of Hawking's quantum cosmology would be first prodded, then ripped and, finally, torn down altogether. In fact, a careful read of *Brief History* shows that this process had already begun in Hawking's mind quite some time beforehand. While circling around the main implications of the no-boundary proposal, for example, Hawking suddenly throws a haymaker: 'If the universe is completely self-contained, with no singularities or boundaries, and completely described by a unified theory, that has profound implications for the role of God as Creator.'[5]

Perhaps he had been riled up by proceedings at a cosmology conference hosted by the Vatican in 1981 – a conference at which he first began to publicly formulate the no-boundary proposal. In *Brief History*, Hawking mentions an address by the then Pope – John Paul II – that seems to have wound him up somewhat:

> [The Pope] told us that it was all right to study the evolution of the universe after the big bang, but that we should not inquire into the big bang itself because that was the moment of Creation and was therefore the work of God. I was glad then that he did not know the subject of the talk I had just given at the conference – the possibility that space-time was finite but had no boundary, which means that it had no beginning, no moment of Creation.[6]

Others, including astrophysicist Frank Tipler,[7] have pointed out that this is not what the Pope said at all. Instead, he was supportive of science and did not tell any scientist to avoid studying certain things. His actual words – that, ironically, were a pretty good assessment of where Hawking's initial singularity had left cosmology – were:

> Every scientific hypothesis about the origin of the world, such as the one that says that there is a basic atom from which the whole of the physical universe is derived, leaves unanswered the problem concerning the *beginning* of the universe. *By itself* science cannot resolve such a question.[8]

Still, with comments like his ones about the Pope flying about, Hawking had set himself on course for a career change. He was, effectively, doing a Pudzianowski; he was diversifying. Yet, just as the Pole had not been any old slugger when he first climbed into the ring, Hawking was hardly any old witterer when he first dipped his toe in philosophy. So how would this all turn out? Could Hawking do better than Pudge?

Is philosophy dead?

Before we begin our analysis, it is perhaps worth double-checking its premise – is Hawking *really* switching disciplines just by weighing in on the Big Questions? Can't his home territory – that of the sciences – tell us why we are here or where we come from? Isn't science all we *need*?

Some of the biggest players in the game think not – they argue that answering existential queries like the ones above is not what science is *for*. This is the line taken, for instance, by quantum-physicist-turned-Anglican-priest John Polkinghorne: 'Science and religion ask different questions of reality: in the former case *how* things happen; in the latter whether there is *meaning, purpose and value* in what is happening – issues that science tends to rule out of its discourse.'[9] Polkinghorne is not rubbishing science, of course – he is not even limiting its reach. His point is that the way it addresses a profound question is not the one people are generally after when they ask it. Scientific answers are pragmatic and functional, says Polkinghorne – and greatly useful in their own right – but they are not enlightening when it boils down to genuine human significance:

> Science and religion, therefore, complement each other, rather than being rivals on the same turf. For full understanding, we need both sets of insights. To take a homely example, the kettle is boiling because gas heats the water (process) *and* because I want to make a cup of tea (purpose).[10]

For Polkinghorne, Hawking's astrophysics can lead to interesting mathematical discoveries about our origins, but it is underpowered on matters of our humanity. This implies that the biggest ideas – the ones folk *really* have in mind when they ask 'who am I?' – still belong with the philosophers and theologians. Hawking, therefore – if he truly wants to make a useful contribution – must indeed set foot on new ground.

The thing is, though, that the Cambridge cosmologist doesn't quite see it like that. In *Brief History*, we begin to see hints that philosophy might not be his bag:

> The people whose business it is to ask *why*, the philosophers, have not been able to keep up with the advance of scientific theories . . . in the nineteenth and twentieth centuries, science became too technical and mathematical for the philosophers.[11]

He goes even further in the (in)famous opening of *Grand Design* – no one, he says, should even be *trying* to answer Big Questions unless they know their way around a formula or three:

How can we understand the world in which we find ourselves? . . . Where did all of this come from? Did the universe need a creator? . . . Traditionally, these are questions for philosophy, but *philosophy is dead*. Philosophy has not kept up with modern developments in science, particularly physics. Scientists have become the bearers of the torch of discovery in our quest for knowledge.[12]

This triumphant declaration is more than a little suggestive of *scientism* – the belief that the only real answers of merit are scientific ones. Anyone dissatisfied with them, therefore, is wishing for more meaning than the universe can ever offer us – the numerical responses of science are final and there simply *is* no deeper layer of understanding.

Whether or not Hawking really means to champion such pure-bred scientism in *Grand Design* is rather doubtful. The book itself is far broader than a regular scientistic tome – his discussion of our cosmic existence is full of imagination, of speculation and of *maybes*. Many commentators have been amused by this, citing the volume to be more of a work of philosophy than it is of science – an irony that is not lost on Oxford mathematician John Lennox:

Apart from the unwarranted hubris of this dismissal of philosophy (a discipline well represented and respected at his own university of Cambridge), it constitutes rather disturbing evidence that at least one scientist, Hawking himself, has not even kept up with philosophy sufficiently to realise that *he himself is engaging in it* throughout his book.[13]

We find ourselves, then, in a rather odd position: Hawking has declared philosophy dead but has then gone on (perhaps unwittingly) to enthusiastically resuscitate it. If we want to trace his thinking, what should we do? Should we walk the path he originally laid out or follow the one he went down? Should we look for our answers in unfeeling calculations or get stuck in to Hawking's highly creative and pseudo-theological thinking?

To philosophize or not to philosophize: that, it would seem, is the question.

You Kant be serious!?

If we genuinely want to get to understand Hawking, it appears we have no choice: we will have to do some philosophy, despite its rumoured death. Our starting point will be the philosophy of science itself – a topic that becomes surprisingly weird in a surprisingly short time. This is because it centres on a basic but strangely elusive problem: how do we know what we *know*?

Taking a step back for a moment, we should remember that science is a truly extraordinary thing: human beings – by asking questions, making observations and doing experiments – have somehow managed to discover many of the laws of nature. For some remarkable reason, our thought processes are so resonant with our universe that we are able to deduce its rules. Yet how, and why, should this be? Einstein, perhaps the best unveiler of our cosmos, was stumped: 'this fact is one which leaves us in awe, but which we shall never understand. One may say "the eternal mystery of the world is its comprehensibility".'[14]

Maybe, as many have argued in the past, this is evidence for a Grand Designer – One whose mind devised both our minds and our universe, thus ensuring a close God-and-Us-and-Nature relationship. This was the conclusion of the famed German astronomer, Johannes Kepler:

For the theatre of the world is so ordered that there exist in it suitable signs by which *human minds, likenesses of God*, are not only invited to study the divine works, from which they may evaluate the Founder's goodness, but *are also assisted in inquiring more deeply.*[15]

In using our scientific success to get on to the God question this early, however, we could be one or two steps ahead of ourselves – for how can we be so sure that our ideas about the world are right in the first place? Sure, we can use Einstein's general relativity to calculate the orbit of Venus; but does that *prove* that gravity is real? Does it *prove* that Venus is moving? How do we know that Venus even *exists*?

At first, this seems nothing but nonsense – it is *obvious* that Venus exists; it is *obvious* that GR is correct. We can see our twin planet with the naked eye, for heaven's sake; we've predicted its motion so successfully that we have sent probes there – we've even taken photos on the surface. How could anyone claim, with any seriousness at all, that Venus might not actually be *real*?

The issue is not so clear-cut, though – as philosophers have been pointing out for centuries. In the end, they remind us, the only information we ever have about anything (Venus included) has come via our sense organs and has been pre-processed by our brain – try as we might, we can never have pure, direct, unmediated access to the world.

Immanuel Kant (1724–1804) made his name by drawing attention to this problem in his mammoth *Critique of Pure Reason* in 1781. Here is an example of his argument:

If we take away the subject [us], or even only . . . our senses in general, then not only the nature and relations of objects in space and time, but even space and time themselves disappear . . . these, as phenomena, cannot exist in themselves, but only in us. *What may be the nature of objects considered as things in themselves and without reference to the receptivity of our sensibility is quite unknown to us.*[16]

In other words, Kant's theory – sometimes referred to as *idealism* or *anti-realism* – claims that we have no idea at all about what reality truly is; all we have are images and sensations within our minds.

Sure, he says, *something* causes an astronomer's brain to paint an internalized picture of a hot rocky ball as she gazes through what gives the overwhelming impression of being a telescope – but that something is not necessarily Venus. It could be *anything* – all the astronomer can ever know is the effect it is having on her mind.

Putting it slightly differently, just because our senses tell us Venus is there doesn't mean it *is* – an entirely different phenomenon could be causing the exact same sensation and we would never have any way of knowing it. The truth – the *real* truth – is for ever inaccessible to us.

We have already seen how Hawking, in the 1960s, hid the beginning of the universe from us – but his work is tame in comparison with Kant's. Two hundred years before Hawking's singularity theorem, the German had taken the whole darn show off the table. We can forget about understanding exotic events like gravitational collapses in space–time, it would seem – Kant says that even something as mundane as rain is an unfathomable mystery:

> Not only are the raindrops mere phenomena, but even their circular form, nay, the space itself through which they fall, is nothing in itself, but both are mere modifications or fundamental dispositions of our sensuous intuition, while *the transcendental object remains for us utterly unknown.*[17]

It is hard to know where to go from here. Idealism doesn't only appear to be opposed to common sense, it also threatens to drive a stake through the heart of science. If reality is wholly unknowable, then aren't scientists – who are, ultimately, trying to describe reality – wasting their time? Not necessarily. At least, not if Hawking has anything to do with it.

Being critical about realism

The ultimate antidote to idealism is *realism*. In its simplest form, *naive realism*, it says that what we see is what we get – if it *looks* like Venus is orbiting the Sun, then that's because there *is* a Venus orbiting a Sun. Most scientists throughout most of history have operated on some sort of realist basis – and it has served them pretty well.

Given Hawking's apparent disdain for philosophy, we might expect him to ignore anti-realists such as Kant completely. Why bother pontificating about whether or not we can access reality when science has been doing precisely that for millennia? Hawking, if he was to take any position at all, would surely choose the no-nonsense school of naive realism – wouldn't he?

The thing is, though, that Hawking is not daft. Although naive realism appeals to our common sense it can, sometimes, be too simple. Over the last century or so, scientists have made a series of increasingly weird discoveries (QM and GR included) that completely defy our first impressions and call that same common sense into question. Hawking is all too aware of this:

> Until the advent of modern physics it was generally thought that all knowledge of the world could be obtained through direct observation, that things are what they seem, as perceived through our senses. But the spectacular success of modern physics, which is based on concepts such as Feynman's [sum over histories] that clash with everyday experience, has shown that that is not the case. The naïve view of reality therefore is not compatible with modern physics.[18]

If Hawking is rejecting naive realism, and if anti-realism appears to remove the point of science, then what is a scientist to do? Is Venus there or not? Has Hawking managed to sit himself down right on the horns of a dilemma?

At this point, the average handle-turning practitioner is past caring; science has put man on the moon – it clearly *works*, so what's the problem? Big Thinker Hawking, however, wants a proper foundation for his theories, so he uses nearly the first third of *Grand Design* to build one – that is a surprising amount of philosophy from someone who considers the discipline dead. So what, exactly, did he come up with?

A rose by any other name

The third chapter of *Grand Design* begins with an account of a goldfish. Because it views the external world through a spherical bowl, the world's straight lines appear curved. Hawking points out that the goldfish, by taking careful measurements, could deal with this distortion – it could still deduce a modified form of Newton's Laws of Motion and apply them to make good predictions. These fishy equations would have to be different from ours to cope with the apparent curvature, yes – but in the end, they would work just as well.

Here, we have two different sets of equations acting as equally successful descriptions of reality – so why, Hawking says, should we have to decide between them? This why-pick-a-side-when-you-don't-have-to idea goes on to form the basis of his own personal philosophy of science – one he labels *model-dependent realism*. In one of many philosophical sections in *The Grand Design*, Hawking explains how his newly developed approach can avoid the idealism–realism hullabaloo and get right to the heart of the matter:

> Model-dependent realism short-circuits all this argument and discussion between the realist and anti-realist schools of thought. According to model-dependent realism, *it is pointless to ask whether a model is real, only whether it agrees with observation.* If there are two models that both agree with

observation, like the goldfish's picture and ours, then *one cannot say that one is more real than another.*[19]

It would seem that Hawking advocates this way of thinking because it is more practically helpful – he is being, essentially, pragmatic. All the scientist really needs to know is whether a given theory will *work*. Any hand-wringing existential debate about whether the theory is *true* is irrelevant – because truth, at the end of the day, doesn't make any significant difference:

One can use whichever model is more convenient in the situation under consideration. For example, if one were inside the bowl, the goldfish's picture would be useful, but for those outside, it would be very awkward to describe events from a distant galaxy in the frame of a bowl on earth.[20]

Hawking, the driven and highly successful physicist, says that the way to get on in scientific life is to concentrate on practicality and results. His version of realism – at first glance, anyway – seems to cut the Gordian knot and bring common sense back into the game. If two or more theories give equally good predictions, why worry about which one is *true*? This is Hawking's rewrite of 'a rose by any other name would smell as sweet', isn't it?

Well, not quite.

One foot in each camp

Niels Bohr (1885–1962) was one of the first theoreticians to ascribe wave-like behaviour to atoms and, as such, was a key player in the heady early days of QM. Like many other cutting-edge physicists, he engaged in more than a little philosophy himself and – according to his friend Aage Petersen – his take on the realism versus anti-realism controversy was an interesting one:

> *Bohr would answer 'There is no quantum world.* There is only
> an abstract quantum physical description. It is wrong to think
> that the task of physics is to find out how nature *is*. Physics
> concerns what *we can say about nature.'*[21]

This is clearly not realism. Bohr does not believe that QM tells us
what the world is really like; instead, it is a tool that helps us make
predictions and build machines. Interestingly, though, his view
seems very similar to Hawking's model-dependent *realism* – so is
Hawking actually a realist or not?

Kant maintained that our knowledge about the universe must
come from the senses; it was, in fact, this very concern that led to his
anti-realist stance. Oddly, though, Hawking seems to agree with him:

> There is no way to remove the observer – us – from our perception
> of the world, which is through our sensory processing and
> through the way we think and reason. Our perception –
> and hence the observations upon which our theories are
> based – is not direct, but rather shaped by a kind of lens, the
> interpretative structure of our human brains.[22]

Pressing on with this line of thought, Hawking eventually arrives at
his personal philosophy-of-science summit – at which point he goes
on to plant a rather anti-realist flag: 'These examples bring us to a
conclusion that will be important in this book: *there is no picture-
or theory-independent concept of reality.'*[23] This is an astonishing
statement. While most scientists would concede that the mathematical weirdness of QM and GR can make us confused about what is
actually going on, it is another thing altogether to state that *science
cannot ever tell us the truth about anything.*

Yet this really does seem to be what Hawking means. He even
gives an example – that of *quarks*, subatomic particles which appear
in the much-maligned but highly accurate standard model:

It is certainly possible that some alien beings with seventeen arms, infrared eyes and a habit of blowing clotted cream out of their ears would make the same experimental observations that we do, but describe them without quarks. Nevertheless, according to model-dependent realism, *quarks exist in a model* that agrees with our observations of how subnuclear particles behave.[24]

Hawking is very, very close here to claiming that quarks *do* exist for us and *don't* exist for the alien. According to his philosophy, asking whether quarks are *real* is a meaningless question – there is no ultimate, model-independent reality to decide the issue. The two competing models are all that there is. If they are equally successful, then they are equally 'real' – or, it would seem, anti-real.

Hawking, it would appear, has not really placed himself on either side of the divide; he seems to have one foot in each camp. He calls his idea model-dependent *realism* and clearly champions the scientific investigation of reality – and yet, at the same time, he agrees with the *anti-realists* about the fundamental unknowability of 'truth'.

What Hawking has really done, then, is blur the lines between realism and anti-realism, utilizing a little bit of both. Can this philosophical blurring, as paradoxical as it would be, somehow make the overall picture clearer? Does it result in the pragmatic science Hawking is seemingly after? How does it actually *work*?

Introducing the multiverse

Turning the clock back to the Copernican revolution (see Chapter 4), Hawking believes that model-dependent realism has an important lesson to teach us. Copernicus, famously, rearranged our solar system, placing the Sun and not the Earth at the centre. For hundreds of years since, scientists have declared this celestial switch to be clearly and indisputably *correct*. Hawking, though, thinks otherwise:

So which is *real*, the Ptolemaic or Copernican system? Although it is not uncommon for people to say that Copernicus proved Ptolemy wrong, that is not true. As in the case of our normal view versus that of the goldfish, *one can use either picture* as a model of the universe . . . the real advantage of the Copernican system is simply that the equations of motion are *much simpler.*[25]

For Hawking, truth is a limited concept. If a model – *any* model – matches observations and makes good predictions, then it is as 'true' as anything ever could be. Sure, we can express a *preference* for a simpler or more elegant theory – but Hawking says that even that call remains basically subjective.

This whole discussion is beginning to sound surprisingly un-scientific. Hawking, though, gets us back on track by reminding us how physics has progressed since the time of Copernicus. When the Earth was shifted from its central throne, some saw it as an object lesson about our insignificance in the cosmos – but modern physics has turned this interpretation on its head. The latest results seem to suggest that the universe really does revolve around *us*:

Observations in the seventeenth century . . . lent weight to the principle that we hold no privileged position in the universe . . . But the discovery relatively recently of the extreme fine-tuning of so many of the laws of nature could lead at least some of us back to the old idea that this grand design is the work of some grand designer.[26]

The 'extreme fine-tuning' that Hawking talks about is genuinely re-markable. Our world has just the right number of dimensions, just the right number of forces, just the right strength of forces, just the right types of particles and just the right set of laws to govern them all. Change anything, by even the tiniest amount imaginable, and the complexity that human beings and life rely on never comes into existence. No wonder he refers to *design*.

Hawking, however, prefers an alternative, non-divine explanation for this fantastical state of affairs – the *multiverse*. When he and Hartle built their universal wave function, they applied Feynman's sum over histories to the cosmos. In doing so, they completely re-imagined its beginning – or perhaps we should say its *beginnings*.

Rather than their universe beginning in the traditional sense, it sort of 'fuzzes' into existence via Feynman's countless paths, as we saw in the previous chapter: 'in this view, the universe appeared spontaneously, *starting off in every possible way*'. The consequence is that the no-boundary wave function actually contains multiple possible universes, each one a variation on ours – it is an infinite ensemble of worlds:

> [The other universes in the HH model] aren't just different in details, such as whether Elvis really did die young or whether turnips are a dessert food, but rather they differ even in their apparent laws of nature. In fact, many universes exist with different sets of physical laws. Some people make a great mystery of this idea, sometimes called the multiverse concept, but these are just different expressions of the Feynman sum over histories.[27]

Given the multiverse, Hawking maintains, the oh-so-perfectly balanced conditions in our one suddenly seem less surprising. The innumerable collection of worlds will include every conceivable combination of laws, particles, forces and events – among this infinite sea, therefore, will sit more than enough universes that are just right for us. The exact fit that we currently enjoy is not only *probable* – the multiverse makes it a *certainty*:

> The fine-tuning in the laws of nature can be explained by the existence of multiple universes. Many people through the ages

have attributed to God the beauty and complexity of nature that in their time seemed to have no scientific explanation . . . the multiverse concept can explain the fine-tuning of physical law without the need for a benevolent creator who made the universe for our benefit.[28]

This is all very well, but Hawking is about to make everything blurry again. His highly individual approach of hybridizing physics with both real and anti-real philosophy now leads him on to one of the most curious discussions of all: whether or not, in his no-boundary proposal, the multiverse even *exists*.

Creating history

Before he launches into his metaphysical analysis of the HH multiverse, Hawking concedes that what he is about to suggest is wholly against the norm: 'The usual assumption in cosmology is that the universe has a *single definite history*. One can use the laws of physics to calculate how this history develops with time.'[29] Hawking's alternative approach is inspired by the way in which Feynman's sum over histories imagines all possible paths being taken before then summing them up: he chooses to think of our universe as having *multiple histories*. Each of these histories is different; each is physically valid. It is the vast set of these potted histories that Hawking then thinks of as the multiverse.

Here, though, Hawking has a problem. His no-boundary proposal, with its infinite multiverse of histories, has a tendency to predict universes that are *not* like ours. In fact, the overwhelming majority of its predictions hold no hope whatsoever for us – his theory favours, by an astronomical margin, a depressingly lifeless cosmos.

Hawking's solution to this dilemma is a fascinating one – rather than work from the origin forwards, as is standard, he suggests we do the *reverse*: 'One shouldn't follow the history of the universe from the bottom up because that assumes there's a single history . . . instead one should trace the histories from the top down, *backwards from the present time*.[30]

According to Hawking, this unconventional tactic allows us to weed out any incorrect predictions by a process he calls *selection*. The way it works is this: we take a measurement here in the present – and then we send it back in time.

Let us take the existence of Venus as an example – for Hawking's multiverse hosts a multiplicity of worlds with no Venus. This is a problem simply solved, for we have looked through our telescopes and *seen* Venus. This observation, Hawking says, means that every history in his wave function without a Venus is *wrong* – so our telescopic discovery travels back through time and *removes them all from the multiverse*:

> This leads to a radically different view of . . . the relation between cause and effect. The histories that contribute to the Feynman sum don't have an independent existence, but depend on what is being measured. *We create history by our observation*, rather than history creating us.[31]

Hawking's historical multiverse, then, seems to be changing all the time. Every time anyone takes a measurement, they remove a set of histories from the pool. Selection, in fact, is Hawking's Big Answer to the question 'Why are we here?' – for it declares, by default, that we *must* be.

Yes, his *physics* predicts vastly more universes without us than with us and renders us incredibly unlikely to exist – but Hawking's effect-and-cause *philosophy* overrules it and assures us we will. Why? Because we have effectively been 'measured' – we are, undeniably, *here*.

Thankfully, our hereness has the relieving effect of erasing all the histories in which we aren't – it is a measurement that echoes backwards through billions of years of multiverse and cuts off every branch that doesn't bear *homo-sapiens* fruit. We exist, dear reader, because we *have* to.

Hawking readily admits that his retrocausative selection is essentially a variation on another well-discussed philosophical idea: the *anthropic principle*. In brief, this principle states that the universe only appears to have been designed for us because it *has* to look like that – for if it didn't, we wouldn't be here to comment on it in the first place.

We shall discuss the relative merits of the anthropic principle in the next chapter, but for now there is one more important question to ask: if Hawking can explain both where we have come from (the no-boundary proposal) and why we are here (selection), what does that mean for *God*?

Hawking and God

Hawking talks a lot about God. In *A Brief History*, the word 'God' or 'gods' turn up, on average, on every fifth page. Initially, as one might expect, he deals briefly with the most simplistic of theologies, in which deities are (or were) merely explanatory placeholders for pre-scientific humans: 'Gradually, however, it must have been noticed that there were certain regularities: the sun always rose in the east and set in the west, whether or not a sacrifice had been made to the sun god.'[32] The same point appears in *Grand Design*, but with a slightly more sardonic tone:

In Viking mythology, Skoll and Hati chase the sun and the moon. When the wolves catch either one, there is an eclipse. When this happens, the people on earth rush to rescue the sun or moon by making as much noise as they can . . . after a time people must have noticed that the sun and moon soon emerged from the eclipse regardless of whether they ran around screaming and banging on things.[33]

Hawking – rightly – thinks that modern physics has dealt pretty comprehensively with the likes of Skoll and Hati and has rendered such childlike beliefs untenable. As a result, he seeks to ramp up the discussion to consider more sophisticated theistic views:

> With the success of scientific theories in describing events, most people have come to believe that God allows the universe to evolve according to a certain set of laws and does not intervene in the universe to break these laws.[34]

Here, though, we have a problem – for this conception of God is not one adhered to by any of the world's major religions. When Hawking writes 'most people', then, it is not entirely clear who he actually means. The quantum cosmologist is at risk of coming across as more than a little out of touch with modern theology and philosophy – the very charge he had laid down in reverse, we remember, at the beginning of *Grand Design*.

As it happens, the God he describes in the extract above sounds more like the God of *deism* – a pseudo-religion that had its heyday among Western intellectuals in the eighteenth and nineteenth centuries. The 'God' of the deists is vague and unknowable – an inscrutable Architect who designed the cosmos, set it going and then hot-footed it off somewhere else. That Hawking has this image in mind is reinforced by his very next sentence: 'However, the laws do not tell us what the universe should have looked like when it started – it would still be up to God to wind up the clockwork and choose how to start it off.'[35]

Having established this picture of God – the one he thinks most people believe in – Hawking can now hammer home the huge significance of his physics. After all, if the only role for God is that of First Cause – a Necessary Being who pushes over domino number one – then why keep Him around if no First Cause is actually needed? Hawking makes this exact point in *Brief History*:

So long as the universe had a beginning, we could suppose it had a creator. But if the universe is really completely self-contained, having no boundary or edge, it would have neither beginning nor end: it would simply be. What place, then, for a creator?[36]

and then repeats it, with a bit more philosophy, in *The Grand Design*:

[In the theistic view] it is accepted that some entity exists that needs no creator, and that entity is God. This is known as the first-cause argument for the existence of God. We claim, however, that it is possible to answer these questions purely within the realm of science, and without invoking any divine beings.[37]

What, though, of the God of the Bible – the God of Polkinghorne, Lennox and Gingerich; of Lemaître, Faraday and Maxwell? Theirs is not some distant vagueness that kick-started the show and disappeared; they believe in a personal God who is heavily involved in human affairs on a day-by-day basis – a God who is hands-on to the point of not only creating, but *visiting* our planet in the person of Jesus.

Such a God gets no mention whatsoever from Hawking – he does not engage with the kind of thoughtful, biblical theism that is actually fairly common among his fellow physicists. Despite the fact that some of Hawking's closest friends and collaborators are Christians, his analysis of 'God' covers animistic myths and Ancient Greek Prime Movers – and then, just like that, it stops.

Switching the spotlight

In this chapter, then, we have watched on as a titan of theoretical physics crossed boldly over into the territories of philosophy and theology; we have seen him introduce model-dependent realism, the historical multiverse and selection. These three ideas, Hawking believes, conspire to tell us where we came from and why we are here – and that, as a result, we have no need of God.

What are we to make of this? Has the great man somehow managed to do what Pudge couldn't and master a whole new discipline? Is God really gone? Is Hawking *right*?

To really answer these questions, we need to dig a little deeper. We need to look, one more time, at Hawking's science – this time through the eyes of his peers. What do they have to say about the current status of his work? What do they affirm? What do they deny?

We will also need to review his philosophy, again from the perspective of the experts. What is their verdict on his model-dependent realism? How do they think his science informs the big ideas of who we are and why we are here? Are these philosophers really out of date, as Hawking says they are? Is science really our only hope of getting to the truth? Is there even a 'truth' in the first place?

In our final chapter, therefore, we shall turn the spotlight around: rather than Hawking analysing the world, the world will analyse Hawking. We have been studying some very Big Questions – might we finally end up, when all has been said and done, with some Big Answers?

8

The Test of Solomon

Solomon – Israel's third King – became known, right across the world, for his great wisdom. Multiple nations at the time (circa 970 BC) sent their brightest delegates to hear him speak and many of his words were carefully recorded for posterity. As a result, we have copies of them available to us even today – and it is one of these sayings, found in the book of Proverbs, that is highly pertinent now:

> The one who states his case first seems right, until the other comes and examines him.[1]

Hawking, living nearly three thousand years after Solomon, has had his turn: he has argued for a singularity at the start of the universe; for the emission of particles from black holes; for a timeless, no-boundary origin of the cosmos; for model-dependent realism; and *against* the existence of God. He has, therefore, stated his case.

Now, in this final chapter, Hawking shall face the Test of Solomon: for it is time for others to come and examine him.

What about singularities?

In Chapter 5, we saw Hawking's PhD make quite a splash. He took the work of Roger Penrose on collapsing black holes and applied it

163

to the expanding universe – in reverse, of course. The result was an unscienceable singularity In The Beginning.

As unpopular as this may be – physicists hate infinities – the idea has stuck around. Find any popular-level book on the topic (or an entry-level textbook for that matter) and there it is: the inscrutable, unknowable singularity, shrouded in mystery, right at the start.

Hawking's brilliant insight, then – as far as the mainstream is concerned – is alive and well. Indeed, in November 2018, a copy of his game-changing thesis sold for a staggering £584,750. And yet, within the halls of quantum cosmology, things are not quite so clear-cut. Does this mean that the buyer should be worried?

In recent years, Hawking's formulation, in its original form, has come under fire. Some of the assumptions he used have been shown to be highly questionable and the outcome is that his version of the singularity proof has quietly died a death. At least, it sort of has – for there are other good pieces of evidence that, broadly speaking, he and/or Penrose might well still be right overall. Why only 'might'?

Well, what we will discover in this chapter (repeatedly) is that modern cosmology is *speculative*. For someone outside of the theoretical physics circle, ideas published in *Nature* or even *The Times* assume a kind of automatic, nailed-on certainty – but science isn't really like that. Cosmological calculations are phenomenally difficult – they tend to require huge simplifications, grand assumptions and an awful lot of guesswork.

Some fairly significant results, especially a famous one from the minds of Alexander Vilenkin, Arvind Borde and Alan Guth,[2] make it seem very likely that our universe *does* have a beginning and that the beginning *is* a singularity. The case is not closed, however – Aron Wall, whom we encountered when discussing the HH universe, sums up the current status of the is-there-a-singularity-at-the-start issue in the following way:

Still, I think that the Penrose theorem is connected to enough other deep principles of physics that *something* like it will probably be true and important in the final theory of physics. Other physicists think that singularities are so disturbing that any 'complete' theory of physics should eliminate them.[3]

This is a wonderful glimpse of cutting-edge quantum cosmology. It is full of opinions, gut instincts and even emotion. This doesn't, of course, make it *wrong* – but it is a highly welcome reminder that quantum cosmologists are people too.

What about Hawking radiation?

If the hand-waving answer from the experts to 'Was Hawking right about the singularity?' is 'We think he might be', can we get something a little more concrete when it comes to black hole radiation?

Not really, no.

Here, the problem with verifying the result is built-in. Hawking's theory says that we can determine the *temperature* of a black hole if we know its *mass*, but the outcome is remarkably frustrating. It turns out that the *bigger* the black hole, the *colder* it is – and this relationship makes everything very messy.

It is much easier, of course, to detect huge black holes, for they affect their surroundings more dramatically. In fact, as this manuscript was being finished, the first ever photograph of a black hole was published – found at the centre of the *Messier 87* galaxy, it has 6.5 billion times the mass of the Sun.

The problem is that these giants, although detectable because of the vast swathes of material being pulled into them, are also very, very cold – far colder even than the frigid space around them. This means we can't use them to pick up Hawking radiation – they simply don't emit enough of it.

Are there any black holes that are hotter than space – hot enough to glow a nice bright orange on a heat-seeking camera? Well, theoretically, yes, there *could* be – but they would be ridiculously tiny and ridiculously short-lived. Small sizes and short lifetimes are bad news for astronomers: hot black holes, if they exist, have proved impossible to find. There is something profoundly sad about this – a feeling that is reinforced by reading what Hawking wrote, just before his death, in 2018: 'People have searched for mini-black holes of this mass, but have so far not found any. This is a pity because, if they had, I would have got a Nobel Prize.'[4]

The scientific community has not given up, however. One alternative approach is to try and build 'black holes' in the lab by creating physical analogies of them, to see if an equivalent of Hawking radiation is emitted. Once again, though, speculation reigns – here, for example, is an account from *Scientific American* of a potential breakthrough by Italian quantum physicist Daniele Faccio:

> Physicists in the field disagree about exactly what the observation means. Ulf Leonhardt of the University of St. Andrews in Scotland, whose group in 2008 proposed the optical method of producing event horizons that Faccio and his colleagues used, says that the new research indeed represents the first observation of Hawking radiation. But others are not as certain.
>
> 'I still need to be convinced that what they are seeing is the analogue of what Hawking found for black holes,' says William Unruh, a physicist at the University of British Columbia.[5]

Despite the lack of a smoking gun, there is pretty much a consensus on the issue: physicists think that Hawking radiation will prove to be *real*. The mathematical theory, with its combination of GR and QM, is so elegant and tidy that the vast majority believe in it, detected or not.

It seems, then, that Hawking's science is standing up to the Test of Solomon reasonably well – he has (kind of) scored two out of two so far. Did he overstep the mark, however, with his outlandish quantum mechanical wave function of the universe?

What about the no-boundary proposal?

Perhaps, before we launch into our analysis of the Hartle–Hawking model, we should allow ourselves a brief refresher of what, exactly, it entails. The first key point is to remember that the universe, according to the Hot Big Bang theory, was once unbelievably *small* – so small, in fact, that the effects of QM should come into play.

Here, we hit a problem: the bread and butter of QM is the wave function, an equation usually only applied to systems of a few particles. If Hawking wanted to write a wave function for the entire universe, he would require some (major) tweaks.

The first of these was the use of *imaginary time* – a decision that rendered Hawking's early universe (if 'early' can really mean anything) *timeless*, with four dimensions of space and none of time. After a while (if 'while' can really mean anything) time switches back to being real again – so the Big Bang may proceed as normal.

The second was to rethink A, the cosmological starting point. Feynman's sum over histories needs an A, some rules and a B. Hawking and Hartle, in their new formulation, ditched the A altogether by wrapping it up into their rules – the universe 'started' in every way, from everywhere, all 'at once'. The joy of this, from Hawking's point of view, is that there is no singularity to be found in the imaginary time regime. There simply are no infinities – no 'boundaries' – and the wave function holds firm all the way through.

The third was to limit themselves to a stripped-down *minisuperspace model* – their idea could cope with a simplified cosmos, but no more. In other words, they were able to establish and test a *technique*, yes – but at the cost of a definitive result.

Where, then, does the no-boundary proposal now stand? What do his colleagues think of it now, nearly four decades after its release? Can Hawking extend his run? Can he make it three out of three?

No rescue

To this day, Hawking's peers remain in awe of him for the sheer audacity of his placing the cosmos into a single box marked 'QM'. His universal wave function is greatly admired, not only for its vision, but also for its execution. Professor Tom Lancaster, author of *Quantum Field Theory for the Gifted Amateur*, considers the no-boundary proposal to be almost a work of art: 'Hawking's earlier technical writing is crystal-clear . . . the Hartle–Hawking paper is beautifully written.'[6] It is all very well for a piece of work to be brave, original, elegant and ground-breaking – but what we really want to know is whether it is *correct*. So is it?

In the end, as Hawking made clear with his pragmatic model-dependent realism, what really matters the most in science is *results* – a theory can only be as good as its predictions. In the years since 'Wave function of the universe' first came out, others have re-run the model in more detail – and, most of the time, it has failed.[7]

One such challenge to the HH universe has come from a genuine heavyweight: Leonard Susskind, Professor of Theoretical Physics at Stanford University. To understand the significance of his attack, though, we will need – albeit briefly – to get an update on the cosmos itself.

Astronomical observations over the last couple of decades have demonstrated to physicists that the universe is not only still expanding, but that the expansion itself is speeding up. This was a shocking discovery to nearly everyone, as scientist-and-author Robert Klauber points out:

> Remarkably, our universe seems to have a positive cosmological constant. This means matter in the universe is being repelled away from other matter, and the universal expansion rate is accelerating. This has flabbergasted the physics community
> 1) as it was totally unexpected, and
> 2) because no one, to this date, knows why this is or what causes it.[8]

Because this result was entirely unanticipated, it led to the re-evaluation of much that came before it. The HH universe is no exception – and, when Susskind applied the new information to it, the outcome was devastating. Hartle and Hawking's no-boundary proposal predicted a universe, yes – but one that contained, give or take, *nothing*.

Despite being rocked by this blow, one of the HH universe's greatest champions swung back with a counter of his own. Don Page, an often-time collaborator with Hawking, tried to rescue the model with a paper that does exactly what it says on the tin: 'Susskind's challenge to the Hartle–Hawking no-boundary proposal and possible resolutions'.

In his dogged determination to save Hawking's brainchild, Page suggests no fewer than *eight* different ways around the issue Susskind had found. First, he states the problem:

> Given the observed cosmic acceleration, Leonard Susskind has presented the following argument against the Hartle–Hawking no-boundary proposal for the quantum state of the universe: It should most likely lead to a nearly empty large de Sitter universe, rather than to early rapid inflation.[9]

However, despite his clear commitment to the HH universe – he declares 'I am loath to give it up' – Page goes on to admit defeat, saying of his eight manoeuvres that he is 'not too happy with any of them'. He finishes up sounding almost like a broken man:

> In summary, Susskind has raised a serious challenge to the Hartle–Hawking no boundary proposal for the quantum state of the universe. There are several potential resolutions of this challenge, but it is not yet clear whether any of them is satisfactory. If no resolutions can be found, the challenge leaves us with the mystery of what the quantum state might be to be consistent with our observations.[10]

Susskind is not the only big-hitter to take on Hawking. Neil Turok, former director of the prestigious Perimeter Institute of Physics, recently co-authored another cosmological onslaught. He doesn't pull his punches, either – the 2018 article, entitled 'No rescue for the no boundary proposal', is occasionally brutal:

> However, since its beginnings, the proposal has suffered from the lack of a precise mathematical formulation.
>
> The no boundary proposal . . . cannot in any way describe the emergence of a realistic cosmology.
>
> Whichever point of view one prefers, the conclusion in all cases is that the no boundary proposal becomes untenable.[11]

It doesn't look good from the HH universe. Some of the top names in the game have found it wanting and even its most ardent supporters – the ingenious Page included – have struggled to defend it.

Not everyone, however, has abandoned the no-boundary ship: the philosopher Quentin Smith, at one point, described the proposal as 'stronger than Darwin's theory of evolution'.[12] Reacting to Smith with incredulity, Aron Wall lets another piece of vital information slip – for it seems that even James Hartle, the co-founder of the model, might be losing faith:

Oh my! Neither Stephen Hawking nor Jim Hartle would make the claim that the Hartle–Hawking state is anywhere near as solidly supported as Darwinian evolution; in fact Jim told me just the other day that *he isn't particularly committed to it being true.*[13]

Leaving the physics aside for a moment, it seems that there may also be a problem with Hartle and Hawking's *maths*. The pair use imaginary time to harness Feynman's technique and dodge the singularity before switching back to real time again – but is this valid? Reed Guy and Robert Deltete don't think so – as they make all too clear in the opening of their withering critique:

Recent models in quantum cosmology make use of the concept of imaginary time. These models all conjecture a join between regions of imaginary time and regions of real time. We examine the model of James Hartle and Stephen Hawking to argue that the various 'no-boundary' attempts to interpret the transition from imaginary to real time in a logically consistent and physically significant way all fail . . . We conclude, therefore, that the notion of 'emerging from imaginary time' is incoherent. A consequence of this conclusion seems to be that *the whole class of cosmological models appealing to imaginary time is thereby refuted.*[14]

Ouch.

Applying the Test of Solomon to Hawking's physics, then, has given us somewhat mixed results.

Is there really a singularity at the beginning of the universe? *Probably.*

Do black holes emit radiation? *Probably.*

Did the universe 'fuzz' into existence, timelessly, out of nothing, without a beginning? *Probably not.*

And, just in case we are thrown by seeing this many 'probablies' in an analysis of science, Aron Wall has a helpful reminder for us: 'This is quantum gravity, so none of us really know what we're talking about!'[15]

What about model-dependent realism?

We shall move on, now, from Hawking's physics to his philosophy. He began *The Grand Design* by attacking modern philosophers – judging them to be so hopelessly unaware of developments within physics that they had effectively abdicated themselves from the Big Questions Throne. Is he right?

The beauty of this particular assertion is that it is directly testable: we simply need to find out if any philosophers have kept up to date with cosmology. Well, as it turns out, plenty have.

Take, for instance, an imposing work from Hawking's own institution: Cambridge University Press's *Universe or Multiverse?* released in 2007. It boasts chapters written by – among others – no less than Smolin, Susskind, Page, Weinberg, Hartle and even Hawking himself. Also finding their rightful place in the very same volume and engaging with precisely the same maths and science are *philosophers* like Nick Bostrom and Robin Collins – and quite a few more.

Such cross-pollination, in fact, is pretty common. The previously mentioned imaginary time paper by Guy and Deltete, for example, appeared in *Synthese* – a philosophy journal.

The fact that Hawking, given he has willingly shared pages with many of them, should dismiss scientifically minded philosophers out of hand as he did is borderline inconceivable. Indeed, it gets worse for him – for it seems that these deep thinkers are not just clued up on his science; some of them have also well and truly got his philosophical number.

One such example is William Lane Craig, who analyses *The Grand Design* in a broadcast entitled 'Stephen Hawking: the anti-philosophy philosopher'.[16] In it, Craig – who has worked with Wall and debated cosmology with the likes of Sean Carroll and Lawrence Krauss – is tickled by Hawking's stance. To him, it is as clear as day that the 'all-new' model-dependent *realism* is nothing more than centuries-old Kantian *anti-realism*, but in pseudo-scientific disguise.

To make his case, Craig picks out a section in the book about differing origin stories. To begin with, Hawking mentions the view that God made the world 'not that long ago': 'That is one possible model, which is favoured by those who maintain that the account given in Genesis is literally true even though the world contains fossil and other evidence that makes it look much older.'[17] Next, Hawking states an alternative: 'One can also have a different model, in which time continues back 13.7 billion years to the big bang.'[18]

Having done this, the Cambridge cosmologist expresses his own belief that the second model – because it fits the data better – is 'more useful'. So far, so good. Then, though – and seemingly out of nowhere – his model-dependent realism kicks in: 'Still, *neither model can be said to be more real than the other.*'[19]

Craig, reading all this, comments that Hawking's philosophy has become a weird hybrid of physics and postmodernism – it seems that anyone is equally welcome to any 'truth'. Hawking doesn't stop there, either: 'We form mental concepts . . . these mental concepts are the only reality we know. There is no model-independent test of reality. It follows that *a well-constructed model creates a reality of its own.*'[20]

Hawking, Craig says, has tied himself in knots. He has already admitted that 'well-constructed' is a wholly subjective term, so the bizarre outcome of his argument is that, for a young-earth-creationist, God *really did* make the world a few thousand years ago – while, for an atheist, there *really is* no God and the universe somehow made itself. Craig, seeking to show the absurdity of all this, applies Hawking's philosophy to cosmology itself:

> This is radical postmodernist anti-realism that says we have no knowledge, really, of anything in the world . . . the model creates its own reality . . . there are different realities for different people . . . for Fred Hoyle, who held the Steady State Model, *the universe really does exist eternally in a steady state*.[21]

Ironically, in fact, model-dependent realism goes on to hang itself: for, if it is correct, then *anything* goes – including the view that model-dependent realism itself is mistaken. The almost-amusing consequence of all this is that if it is *right*, it is also *wrong*.

Hawking's unusual thinking, then, fails the Test of Solomon – it doesn't stand up to expert scrutiny. Model-dependent realism dies the same death as other postmodern claims – for all essentially declare it to be 'a truth' that 'there are no truths'. His cosmological variation on a theme, despite its scientific context, is no different – his philosophy carefully saws off the very branch that it sits on.

What about the multiverse?

Hawking's dabble with anti-realism may well have hit the self-destruct button; but is the same true of his multiverse? We saw, in the previous chapter, how his universal wave function plays host to an infinite number of possible universes, each one a history in Feynman's sum. Although these differing universes are nearly all uninhabitable, our just-right cosmos will nevertheless emerge, on its own, from the throng – because our current observations fly back through time, removing countless unsuitable histories.

If, however, only *one* universe actually survives, can we *really* call this a multiverse? Tom Lancaster offers a wise word of caution: 'The sum over histories technique does not commit any-one to a realist stance – one in which they must believe that these histories actually *happened* – it can simply be viewed as a math-ematical scheme.'[22] Lancaster also points out that there is no scientific link between the HH universe and ideas about back-wards-in-time selection: 'Hawking's physics does not require retrocausality.'[23]

Perhaps, though, Hawking's motivation in applying retrocausa-tive thinking lies elsewhere. It is not needed within his proposal, but it does serve another purpose. After all, explaining why our universe should be life-permitting against all the odds is one of an increasing number of holy grails out there for quest-minded physicists.

Hawking, it seems, thinks he can solve this fine-tuning problem with his selection principle. Because, in his scheme, it is our observations that select the 'correct' history, we are bound to exist. Since we can only ever measure cosmological values that allow for us, his one surviving world must therefore be spot-on for human life. He doesn't need some sort of supernatural Designer: goodbye, God.

This is all well and good, but Hawking's no-boundary proposal, on which his selection principle depends, is ropey at very best – as we've seen, it makes incorrect predictions and involves highly question-able maths. When we also consider that the existence of its multiverse is unclear and Hawking's controversial back-through-time selec-tion is physically unnecessary, his explanation of fine-tuning doesn't really seem to hold up.

Does this failure dismiss the multiverse altogether? Is God, the Great Architect, needed after all? Not necessarily – for, as it turns out, there is more than one way to skin an infinite ensemble of cats.

In a key section of his exhaustive *Universes*, John Leslie – a philosopher, by the way – analyses a surprisingly long list of alternative multiverse-making models:

> A multiplicity of Worlds, small-u universes, could help to make any fine tuning unmysterious. The chapter expands the earlier quick survey of mechanisms by which multiple universes might be generated: oscillations, quantum World-splittings, quantum fluctuations, symmetry breakings which set up gigantic zones inside a perhaps infinitely large cosmos, and so on.[24]

In other words, even if Hawking and Hartle are wrong – as we have seen they probably are – we may still be living in a multiverse of some sort or another. Leslie is of the opinion that, because there are so many ways to get multiple worlds, there is probably something to the notion. This is, by no means, the consensus – many physicists pooh-pooh the idea – but it is certainly true that more and more people in and around cosmology are persuaded that some kind of multiverse is realistic.

Hawking's argument about an unnecessary Designer, then, might still have to be dealt with. If there are countless universes – however they may have been made – then isn't it true that at least one will end up being a suitable home for us? Doesn't that, in turn, explain our existence – without any recourse to God?

The multiverse-explains-fine-tuning argument is closely related to the *Anthropic Principle*. Introduced in its modern form by Brandon Carter in 1973 (at a symposium celebrating 500 years of Copernicus), it flirts with being a truism. The basic idea is this: the universe's key numbers (force strengths, dimensions, etc.) *must* be just right for us, because if they were different we wouldn't be here – and so we couldn't measure them.

The anthropic principle can be used as a direct challenge to the theistic *Argument from Design*. Rather than attributing the remarkable coincidences found in every scale of our cosmos to a Creator, they can be dismissed – there *is* no design. Instead, they are necessary truths: human beings, by default, will see a world 'made' for them. And, if there is no design, then there is no longer a design-based argument for God.

At first glance, this assertion seems pretty conclusive. There don't appear to be any gaping logical holes or obvious missteps. Have we hit the end of the theological road?

Professor John Lennox, who we saw was irked by Hawking's criticism of philosophy, thinks not – for he believes, wholeheartedly, in God. How can he justify such a belief in the light of the multiverse? Hasn't the straight thinking of the anthropic principle put an end to all that silly stone-age nonsense? Lennox says no – and gives two reasons straight off the bat:

> Hawking has once again fallen into the trap of offering false alternatives . . . as philosophers have pointed out, God could create as many universes as he pleases. The multiverse concept *of itself* does not and cannot rule God out . . . [and] what of the multiverse itself? Is *it* fine tuned? If it is, then Hawking is back where he started.[25]

The first of these points is interesting indeed: why should anyone believe that God created *one* universe, not many? Don Page, for example – the friend of Hawking's who tried, and failed, to save the no-boundary proposal from Susskind – lists himself as one of a growing number of theistic thinkers who believe that God might just have gone big:

> Leslie, Barr, Collins, Cleaver, Kraay, and others claim that since God is infinitely creative, it makes sense to say that He might create a physical reality much larger than the single visible part of the universe or multiverse that we can observe directly.[26]

This is similar to situations in which some scientists – including Hoyle and Hawking – asserted that a universe without a beginning in time would have no need of God. The response came back from the theists that, even as far back as Augustine in AD 400, many theologians considered God to be somehow above and beyond our idea of time – and were therefore quite content with the notion that He had created a universe that had existed, from our perspective, eternally. As Wall says: 'An Author stands outside the time-stream of their own story.'[27]

God, therefore, survives both arguments about beginnings and the philosophical challenge of the multiverse. In fact, in the case of the latter, He might even prove to *benefit* from it.

This is because, in contemporary physics, Lennox's second point comes into play – for, of the variety of multiverses that are currently on the table, nearly all require their own form of fine-tuning before they can blossom. Anyone wanting to use the Argument for Design could point to this fine-tuning for help – and, just like that, God is back.

Before we move on from the peculiarities of the multiverse and the anthropic principle, however, there remains one more point to be made. For, as unlikely as it may seem, the two ideas are a potential threat to *science*.

Infinite death

A multiverse, officially, is a collection of separate universes – as in, *completely* separate. No information can pass between them; no science experiment can be done 'here' to tell us about 'there'. This has led to many thinkers classing the multiverse as *metaphysics* rather than physics – placing it squarely in the realm of *philosophy*.

So serious is this issue that astrophysicist Luke Barnes, a specialist in fine-tuning, feels he needs to warn fellow scientists who want to jump on to the multiverse bandwagon(s):

We cannot observe any of the properties of a multiverse as they have no causal effect on our universe. We could be completely wrong about everything we believe about these other universes and no observation could correct us. The information is not here. The history of science has repeatedly taught us that experimental testing is not an optional extra.

The hypothesis that a multiverse actually exists will always be untestable.[28]

If multiverse ideas are, by definition, unverifiable, then the ramifications of this are huge. Relying on the multiverse to give us any answers to any of the Big Questions simply won't work – we will be trapped for ever in speculation with no way to confirm our hunches. And, if this wasn't already bad enough, a *real* multiverse – if it was an infinite one – could just about finish science off.

To see how, we can imagine conducting a rather straightforward experiment – measuring the boiling point of pure water. Since water is actually made up of billions and billions of molecules, the temperature at which our sample boils is determined by the *average* behaviour across all of them. This fact actually permits a *range* of *possible* results – of which our usual observation of 100 °C is by far the most probable. In the multiverse, however, we have a severe boiling-point problem.

In universe number one – *U1* – scientists find that their water samples boil at or very near 100 °C every time. They conclude that the boiling point of water is 100 °C. That is fine. In *U2*, however, things are different. This is because, in an infinite multiverse, *every possibility exists*. The lab-workers in *U2* find that the temperature their water boils at goes up by precisely 0.2 °C every third Monday – and this has happened, consistently, since their records began.

The textbooks in *U2*, therefore, carefully explain how the boiling point of water can be predicted for any date in the future – they even include calendars to help engineers know what will happen.

And yet, sooner or later, these predictions – and the machines based on them – will fail.

This is because, unbeknownst to the physicists in them, the *laws of physics in U1 and U2 are actually the same*, despite their *experimental results being different*. This arises because both sets of data are *physically possible* under those same laws – even though those in *U2* are astonishingly improbable – and so *both are bound to occur somewhere* in the multiverse.

In short, across an infinite multiverse, an infinite number of experimentalists will be infinitely misled an infinite amount of times, leading to an infinite number of failed predictions and an infinite number of deaths. Oh dear.

The consequences for our here-and-now science are devastating: for we can have no idea whether (a) we have found genuine laws of nature that will predict our futures or (b) we just happen to have observed a freakish run of very unlikely outcomes for a very long time that will suddenly cease without warning and revert to another principle altogether. Worrying.

The multiverse's partner in crime, the anthropic principle, is also scientifically limited – for it answers the question 'Why is that like that?' with 'because it has to be'. This form of statement can hardly be called helpful, as Smolin reminds us: 'Scenarios invoking the anthropic principle as an explanation for our universe's laws and initial conditions have yet to yield a single falsifiable prediction for a currently doable experiment. I doubt they ever will.'[29] Anyone seeking to call on the multiverse and the anthropic principle to get rid of God, then, had better be careful – for they might end up throwing out the scientific baby with the (holy) bathwater.

What about God?

God, it would seem, is holding up OK thus far. Of course, according to Hawking's own philosophy, he should never have been doubted in the first place: for, if Hawking really means what he appears to say about model-dependent realism, God *must* exist – for anyone who chooses to believe in him. James Clerk Maxwell, Hawking's great unificatory forebear, for example, is firmly on record as a Christian: 'I believe, with the Westminster Divines and their predecessors *ad infinitum* that "Man's chief end is to glorify God and to enjoy Him for ever."' Hawking, to be consistent, should therefore concede that the God of the Westminster Confession is *real* – for Maxwell, at the very least.

It is hard to believe, however, that this is truly what Hawking meant. Instead, the overwhelming sense from his body of work is that he believes the God question is not relative at all – it is a genuine one, with a proper Yes-or-No answer.

The problem, really, is that Hawking does not go deep enough with his theology. For him, God is merely a simple hypothesis who exists only to explain the presence of our cosmos. For the Christian, though, God is far more than that. The Christian's God is neither a vacant landlord nor a vague creating force of some kind; he is an attentive and involved Saviour, intertwined with real human history – as we read here in the New Testament book of Hebrews:

> In the past God spoke to our ancestors through the prophets at many times and in various ways, but in these last days he has spoken to us by his Son [Jesus], whom he appointed heir of all things, and through whom *also he made the universe.* The Son is the radiance of God's glory and the exact representation of his being, sustaining all things by his powerful word. After he had provided purification for sins, he sat down at the right hand of the Majesty in heaven.[31]

Note that 'made the universe' – the totality of Hawking's role for God – is reduced to a mere 'also' in this passage. The Christian message is a whole lot richer than 'God created the world'. It says that we were designed to be in a relationship with God Himself; that we have catastrophically rebelled; that the God-and-man Jesus took our punishment; that He rose again to give us hope of eternal life; that love, forgiveness and full reconciliation are on offer from On High to anyone who seeks Him.

The danger here, of course, is that it seems we are heading down a route every bit as unscientific as the multiverse or Hawking's postmodernism. How on earth are we to decide whether Jesus is really the Son of God? What astronomical observation or mathematical theory could help us out here? Aren't we on looser ground than we were with even the most tentative of QM interpretations or multiverse-generating mechanisms?

Well, not necessarily.

Different universes

Professor Stephen Hawking spent the vast majority of his life searching for the *best-fitting explanation* of our universe. He outlines his core principles in *The Grand Design*. A model is good if it:

1. Is elegant
2. Contains few arbitrary or adjustable elements
3. Agrees with and explains all existing observations
4. Makes detailed predictions about future observations that can disprove or falsify the model if they are not borne out.[32]

It is clear, then, that *observations* are vital in this search. Examples could include:

A: There are three dimensions of space and one of time.
B: Space is very, very big.

C: Space is expanding, and the expansion rate is increasing.

D: There are four distinguishable forces in nature.

E: There are specific types of particles with specific properties.

and so on.

Given all this data, the challenge is to put together a model that satisfies Hawking's criteria. Many have tried this, resulting in many competing and speculative cosmological systems. What is interesting is that a scientist like Hawking can look at the universe *ABCDE* (as we shall call it) and disagree with a scientist like Page about whether or not it is the work of *God*. How can this be?

The answer – and it is a profound one – is this: *they are NOT seeking to explain the same universe.*

Sure, the two friends agree on observations A, B, C, D and E. But the universe doesn't stop there – these are not the only observations about our world. Let us look, once again, at that extract from Hebrews:

> In the past God spoke to our ancestors through the prophets at many times and in various ways, but in these last days he has spoken to us by his Son [Jesus], whom he appointed heir of all things, and through whom also he made the universe.[33]

From these verses, and from others in the Bible, we come across some more potential data:

F: God spoke to the Israelites, through the prophets, predicting events centuries in advance.

G: God Himself came to Earth, in the Person of His Son, Jesus Christ.

H: Jesus performed miracles, including healings.

I: Jesus rose from the dead, three days after being executed by the Romans.

Then, from within the Christian faith, comes yet more:

J: Miracles are still performed to this day, in the Name of Jesus.
K: Lives are miraculously transformed when people declare a faith in Jesus as God.
L: Christians actually experience a personal and real relationship with the Creator God.

Of course, for the atheist – for Hawking, for example – none of these observations apply, because he does not believe them. For him, the data reads:

M: God did not speak to the Israelites.
N: God did not come to Earth.
O: Jesus did not perform miracles.
P: Jesus did not rise from the dead.
Q: Miracles do not happen.
R: Lives are not miraculously changed.
S: There is no relationship to be had with God.

When Hawking and Page[34] seek to explain the 'universe', then, they have two different ideas in mind. Yes, they agree on the *ABCDE* part, but after that, their worlds diverge. Page is looking for the best explanation of *ABCDEFGHIJKL*; Hawking is trying to account for *ABCDEMNOPQRS*.

In their purely *cosmological* work, of course, they will only refer to *ABCDE* – which is common ground. This, unintentionally, creates the illusion that both are working on the *same* cosmos. It is, however, the 'hidden' data – the second half of their lists – that is decisive for them as individuals and leads to their universe-explaining world views.

It might seem, at first, that this argument takes the same form as Hawking's model-dependent realism – different people will arrive at different 'truths' and nothing can then be said or done to further distinguish them. That conclusion, though, would do both Page and Hawking a disservice.

The observations from the second halves are not 'true-for-you-but-not-for-me' statements – they are real-world claims that are either true or false. Neither are they 'untestable', for they are at least partially verifiable assertions about earthly history and about real lives now. We might be living in universe *ABCDEFGHIJKL*, for which the best explanation is Christianity; *or* we might be living in *ABCDEMNOPQRS*, for which the best explanation is currently – and perhaps for ever – unknown.

We are not, however, living in *both*.

Big Answers

As we draw to a close it is worth highlighting, once again, just what an extraordinary man Hawking is. In fact, why not let Page do it?

Stephen Hawking is a remarkable scientist. Despite being afflicted with amyotrophic lateral sclerosis, he has become the greatest gravitational theorist since Albert Einstein. For example, in work with Roger Penrose, Hawking showed that Einstein's classical theory of gravity, general relativity, implied the universe had a beginning. He went on to show that black holes normally can only grow in area, though in 1973 he made the surprising discovery that quantum-mechanical effects can allow black holes to emit radiation and thereby shrink. Since that time, Hawking has focused on applying quantum mechanics to gravity and to the universe as a whole.[34]

As a physicist, then, Hawking is undoubtedly a genuinely brilliant thinker, truly worthy of being hailed as generation-defining. His imagination and invention are close to being unparalleled and he shall never be forgotten. His forays into the territories of philosophy and theology, despite their limitations, made worldwide news and their impact has been huge – the internet remains rife with his increasingly confident dismissal of God.

Hawking's cosmological adventuring took him to the exotic island of the no-boundary proposal – and it was this that he felt removed the need for a Creator. When we applied the Test of Solomon to the HH universe, though, we found that it struggled to stand up to scrutiny. The verdict from physicists and philosophers was that it is highly speculative, probably wrong and does not actually get rid of God even if it is right.

So what, then, of the Big Questions Hawking promised to resolve? Where did we come from? Why are we here?

Ultimately, the answers to these hinge on another query – one that needs sorting out first. It is this: *which universe are we actually in*?

The reason this matters so very much is that cosmology alone – *ABCDE* – does not give us enough to go on. We need to know *more*. In and among the physics of GR, QM, black holes and the multiverse, we also need to know whether Jesus came back to life or not. Here is Page again:

> I personally think it might be a theological mistake to look for fine tuning as a sign of the existence of God . . . In other words, I regard the death and resurrection of Jesus as the sign given to us that He is indeed the Son of God and Saviour He claimed to be, rather than needing signs from fine tuning.[35]

As it happens, it is the conviction of many in the world of physics and of philosophy that Jesus *is* 'the Son of God and Saviour He claimed to be' – the present authors included. This places us in universe *ABCDEFGHIJKL* – for which the best explanation is *God*.

The answer, then, to 'Where did we come from?' is that *Jesus made us* – or, at least, such is the carefully considered conclusion of Page, Wall, Craig, Lemaître, Faraday, Kepler, Copernicus, Michell, Lennox, Lewis, Gingerich, Barnes, Polkinghorne and more. The answer to 'Why are we here?' is, unsurprisingly, also related to Jesus – for all would join Maxwell, and others *ad infinitum*, by saying: 'Man's chief end is to glorify God and to enjoy Him for ever.'

Sadly, then, despite a wonderful career in which he tackled more than a few Big Questions, Hawking never seemed to spend any time on what might just turn out to be the Biggest Of Them All:

'But what about you?' Jesus asked. '*Who do you say I am?*'[36]

Appendix 1
A note on M-theory

While M-theory gets no mentions whatsoever in *Brief History*, it crops up several times in *Grand Design*. In fact, M-theory is Hawking's favoured candidate for a theory of everything:

> M-Theory is the *only* candidate for a complete theory of the universe. If it is finite – and this has yet to be proved – it will be a model of a universe which creates itself. We must be part of this universe, because there is no other consistent model. M-Theory is the unified theory that Einstein was hoping to find.[1]

This paragraph seems to suggest that the holy-grail-of-physics hunt is over; and yet M-theory has not made a single appearance in our main text – *why*?

Well, because M-theory – or *string theory* as it is more widely known – is still very much in its embryonic stages. What's more, its gestation period is clearly a long one, for theoretical physicists have been working on it for decades.

String theory, in broad brushstrokes, says that all the fundamental particles of the standard model are actually just different vibrations of tiny, subatomic 'strings'. The maths can (sort of) account for all of them and, magically, *can also account for gravity.* This second point is precisely why Hawking and so many others champion it.

There are, however, a number of significant problems with string theory. First, it is incomplete – no one knows how to write down the equations fully, let alone solve them. Second, it predicts the existence of *eight* more dimensions of space on top of our familiar three, for which there is no physical evidence. Third, it comes in so many versions that they are pretty much countless and nearly all these versions give universes nothing like ours – if they even give a universe at all. In other words, string theory fails a lot.

This, unsurprisingly, has turned some theoreticians off it in a big way and they have said as much. Lee Smolin, for example, is not a fan: 'We understand very little about most of these string theories. And of the small number we understand in any detail, every single one disagrees with the present experimental data, usually in at least two ways.'[2] And neither is Roger Penrose, whose singularity theorems were so important to Hawking's PhD: '[*The Grand Design*] is a bit misleading. It gives you this impression of a theory that is going to explain everything; it's nothing of the sort. It's not even a theory.'[3]

We shouldn't take these comments as being final, though: M-theory/string theory may well turn out to be correct one day – it might just be the theory of everything that Hawking and his colleagues have been so desperate to find. For that to happen, however, there will have to be huge changes to the current state of affairs. Physicists must find those extra dimensions in reality; they must explain why only one particular version of string theory actually manifested; and they must both write and solve its equations.

Since we are nowhere near any of those things happening, Hawking could perhaps be described as being over-enthusiastic in suggesting that M-theory is the *only* candidate available or in hinting that we are close to getting over the line. History tells us that whenever claims like this are made in the world of physics – as they were by some fairly big names before QM and GR appeared and made them look rather silly – the universe seems to then offer up a new and unexpected surprise.

If he is over-enthusiastic on the possible veracity of M-theory, though, there is another point that Hawking makes in the passage above that is just plain wrong: for a *universe cannot create itself.* To see why, we will need another appendix.

Appendix 2
A universe from nothing

It has become quite the fashionable thing of late for scientists to write popular books that claim our universe could have arisen, by itself, from *nothing*. Lawrence Krauss, Peter Atkins and others have thrown this idea into the public arena in recent years and there is now a growing impression among the laity that this extraordinary event has, indeed, *actually happened*.

It hasn't.

All these assertions, whatever their flavour, have one important thing in common: they are not really universes from nothing at all. Here is Atkins: 'I would like you to think of nothing in a very commonsensically primitive way . . . think of miles and miles of uniform, empty space, and of years and years of time'.[1] And now Krauss: 'I want to be clear about the kind of "nothing" I am discussing . . . I will assume space exists with nothing at all in it, and that the laws of physics also exist.'[2]

In each case the cosmos arises from *something*. For Krauss, the vehicle turns out to be the virtual particles caused by the uncertainty principle. Atkins majors on the principle of the conservation of energy. When Hawking tries this trick – as we shall see – he says that the Laws of Nature themselves gave birth to the universe.

In other words, all these universes that are supposedly from 'nothing' are actually dependent on *something being there already*. The reason that some scientists aren't bothered by this is that their definition of nothing is actually quite different from that of the philosophers. The scientific 'nothing' could mean no mass or no energy or no charge – the philosophers' nothing, on the other hand, means *nothing*.

David Albert, a philosopher who specializes in QM, points out this important difference in a scathing *New York Times* review of Krauss's book *A Universe from Nothing*:

> the fact that particles can pop in and out of existence, over time, as those fields rearrange themselves, is not a whit more mysterious than the fact that fists can pop in and out of existence, over time, as my fingers rearrange themselves. And none of these poppings – if you look at them aright – amount to anything even remotely in the neighborhood of a creation from nothing.[3]

It is not just philosophers who find themselves part-frustrated-and-part-amused at this bizarre redefinition of nothing. Luke Barnes, a cosmologist, also takes issue with Krauss, especially for his strange use of Hawking radiation:

> Inconsistency with 'nothing' abounds. Having admitted that it would be 'disingenuous to suggest that empty space endowed with energy . . . is really nothing', just a few pages later [Krauss] is telling us that in a universe emptied by expansion 'nothingness would reign supreme', and that the creation of particles from the empty space around a black hole shows that 'under the right conditions, not only can nothing become something, it is required to'. The book descends into the ridiculous.[4]

When Hawking says, then, that 'Because there is a law like gravity, the universe can and will create itself from nothing',[5] he is openly contradicting himself. If the law of gravity is there at the start, then it is not creation out of nothing.

There is a recurring theme underlying all this – the desire to have the universe genuinely come out of nowhere is usually espoused in this way by scientists who wish to link the claim, directly, to atheism. Richard Dawkins, perhaps the arch-atheist, can barely hide his glee at Krauss's book:

> Do you think some agent must have caused everything to start? . . . Read Steven Weinberg, Peter Atkins, Martin Rees, Stephen Hawking. And now we can read Lawrence Krauss for what looks to me like the knockout blow. Even the last remaining trump card of the theologian 'Why is there something rather than nothing?' shrivels up before your eyes as you read these pages.[6]

Hawking, then, makes it into Dawkins' you-don't-need-to-believe-in-God-any-more list. This is a great pity, because he has had so much to say of interest on cosmology, philosophy and theology – but here, just like Krauss and Atkins, he has pushed the philosophical boat out too far.

Physics, without admitting a transcendent Creator God, emphatically does *not* allow for a universe from nothing and – because its models are *always* dependent on an already-present law, principle or the like – *it never will.*

Notes

1

1 Perhaps the best of the volumes it can be found in is Arthur C. Clarke, *The Collected Stories* (London: Gollancz, 2001), pp. 417–22.

2 You can download it yourself from <www.repository.cam.ac.uk/handle/1810/251038>!!

3 Stephen Hawking, *A Brief History of Time* (London: Bantam, 1996), Foreword.

4 *Desert Island Discs*, BBC Radio 4, first broadcast 27 December 1992.

5 Bernard Levin's obituary, *The Times*, 9 August 2004.

6 Hawking, *A Brief History of Time*, Foreword.

7 Stephen Hawking and Leonard Mlodinow, *The Grand Design* (London: Bantam, 2010), Foreword, emphasis added.

2

1 *The Observer*, 18 September 1938.

2 Duncan Steel, 'Tunguska at 100', *Nature* 453 (2008), pp. 1157–9.

3 Stephen Hawking and Leonard Mlodinow, *The Grand Design* (London: Bantam, 2010), p. 24.

4 Christopher Shields, 'Aristotle', in Edward N. Zalta (ed.), *The Stanford Encyclopedia of Philosophy* (Winter 2016 Edition), <https://plato.stanford.edu/archives/win2016/entries/aristotle>.

5 Isaac Newton and Robert Hooke, *Isaac Newton Letter to Robert Hooke*, 5 February 1675, <https://digitallibrary.hsp.org/index.php/Detail/objects/9792>.

6 Stephen Hawking, *On the Shoulders of Giants* (Philadelphia, PA: Running Press, 2002), p. 728.

7 Hawking, *On the Shoulders of Giants*, p. 729.

8 Isaac Newton, *General Scholium*, *Philosophiae Naturalis Principia Mathematica*, <www.isaacnewton.ca/gen_scholium/scholium.htm>.

9 Letter to Richard Bentley, 25 February 1692/3, <www.newtonproject.sussex.ac.uk/view/texts/normalized/THEM00258>.

10 The authors are grateful to Keesing for agreeing to an interview and allowing us to quote his comments here.

11 Hawking, *On the Shoulders of Giants*, p. 1161.

12 Yehuda Elkana, 'Einstein and God', in Peter L. Galison et al. (ed.), *Einstein in the 21st Century* (Princeton, NJ: Princeton University Press, 2008), p. 42.

13 *The Science Book* (London: Dorling Kindersley, 2014), p. 185.

14 For a full description of why mass bends space–time (which assumes no prior scientific knowledge and uses no equations), see Chapter 5, David Hutchings and Tom McLeish, *Let There Be Science* (Oxford: Lion Hudson, 2017).

15 *The Homiletic Review*, April 1896, p. 442.

3

1 Stephen Hawking (ed.), *A Stubbornly Persistent Illusion: The essential scientific works of Albert Einstein* (Philadelphia, PA: Running Press), p. 310.

2 Stephen Hawking and Leonard Mlodinow, *The Grand Design* (London: Bantam, 2010), p. 140.

3 You can find his excellent video on this at: <www.youtube.com/watch?v=p-MNSLsjjdo>.

4 Richard Feynman, *The Character of Physical Law* (Cambridge, MA: MIT Press, 1965), p. 129.

5 Maximilian Schlosshauer et al., 'A snapshot of foundational attitudes toward quantum mechanics', *Studies in History and Philosophy of Science Part B Studies in History and Philosophy of Modern Physics* (2013) 44(3), pp. 222–30.

6 Sean M. Carroll, 'The most embarrassing graph in modern physics' (blog post), 17 January 2013, <www.preposterousuniverse.com/blog/2013/01/17/the-most-embarrassing-graph-in-modern-physics>.

7 Max Planck, 'Religion und Naturwissenschaft', lecture given in Berlin, 1937. This translation from *Scientific Autobiography and Other Papers*, tr. F. Gaynor (New York: Philosophical Library, 1949), p. 97.

8 Werner Heisenberg, 'Reality and its order', in *Collected Works Section C: Philosophical and popular writings: Volume I: Physics and cognition: 1927–1955*, tr. M. B. Rumscheidt and N. Lukens (Munich: Piper, 1984).

4

1 See <www.shanghairanking.com>.

2 Owen Gingerich, *God's Planet* (Cambridge, MA: Harvard University Press, 2014), pp. 12, 13.

3 Nikolaus Copernicus, *Dedication of the Revolutions of the Heavenly Bodies to Pope Paul III* (1543).

4 Arthur Koestler, *The Sleepwalkers* (New York: Macmillan, 1959), p. 191.

5 Gingerich, *God's Planet*, p. 19.

6 Albert Einstein, 'Do gravitational fields play an essential role in the structure of the elementary particles of matter?', in *The Collected Papers of Albert Einstein: Volume 7: The Berlin Years: Writings: 1918–1921*, tr. Alfred Engel (Princeton, NJ: Princeton University Press), p. 83.

7 Simon Singh, *Big Bang* (New York: Harper Perennial, 2005), p. 156.

8 A. L. Berger (ed.), *The Big Bang and Georges Lemaître* (Dordrecht: Reidel, 1984), p. 370.

9 C. S. Lewis, *The Discarded Image* (Cambridge: Cambridge University Press, 1964), p. 16, emphasis added.

10 Singh, *Big Bang*, p. 354.

11 George Gamow, *Mr Tompkins in Paperback* (Cambridge: Cambridge University Press, 2012).

12 Stephen Hawking, *A Brief History of Time* (London: Bantam, 2011), p. 9.

13 David Wilkinson, *God, Time & Stephen Hawking* (London: Monarch, 2001), p. 56.

14 Hawking, *A Brief History of Time*, p. 49.

15 C. S. Lewis, 'Is Theology Poetry?', presented at the Socratic Club, Oxford, 1944. Reprinted in *They Asked For a Paper* (London: Geoffrey Bles, 1962), pp. 150–65, quote from p. 165.

16 Owen Gingerich, *God's Universe* (Cambridge, MA: Harvard University Press, 2006), p. 40.

5

1 *Norm Macdonald Live*, Season 3, Episode 13 (3 October 2017).

2 Rachel Cooke, 'Val goes out on a limb', *The Observer Film* (9 May 2004).

3 Yes, this is a real thing. See <www.webcamtaxi.com/en/spain/ las-palmas/lanzarote-airport.html>.

4 George Musser, *The Complete Idiot's Guide to String Theory* (New York: Alpha, 2008), p. 88.

5 John Michell, 'On the means of discovering the distance, magnitude &c. of the fixed stars, in consequence of the diminution of velocity of light, in case of such a diminution should be found to take place in any of them, and such other data should be procured from observations, as would be further necessary for that purpose', *Philosophical Transactions of the Royal Society*, Vol. 74 (1784), p. 42, emphasis added. See: <https://royalsocietypublishing.org/doi/pdf/10.1098/ rstl.1784.0008>.

6 *Two Infinities and Beyond Part 2*, from *The Curious Cases of Rutherford and Fry*, Series 12, BBC Radio 4, first broadcast 12 December 2018. [He actually said the 'bleep'.]

7 Stephen Hawking (ed.), *Three Hundred Years of Gravitation* (Cambridge: Cambridge University Press, 2008), p. 226.

8 Marcia Bartusiak, *Black Hole* (New Haven, CT: Yale University Press, 2015), p. 80.

9 Albert Einstein, 'On a stationary system with spherical symmetry consisting of many gravitating masses', *The Annals of Mathematics*, Second Series (October 1939) 40(4), pp. 922–36.

10 Stephen Hawking, 'The Properties of Expanding Universes', PhD thesis (1966), Introduction.

11 Brian Greene, *The Elegant Universe* (London: Vintage, 2000), p. 129.

12 Lee Smolin, *The Trouble with Physics* (London: Penguin, 2008), p. 54.

13 Sean Carroll, *The Trouble with Physics* (Review), Preposterous Universe Blog (3 October 2006), <www.preposterousuniverse. com/blog/2006/10/03/the-trouble-with-physics>.

6 Part 1

1 George A. Tramountanas, 'Akira Yoshida: A bullet for Marvel's young guns', CBR.com, 31 March 2005, <www.cbr.com/ akira-yoshida-a-bullet-for-marvels-young-guns>.

2 Stephen Hawking and Leonard Mlodinow, *The Grand Design* (London: Bantam, 2010), p. 90, emphasis added.

3 Hawking and Mlodinow, *The Grand Design*, p. 102.

4 Hawking and Mlodinow, *The Grand Design*, p. 107.

5 Hawking and Mlodinow, *The Grand Design*, p. 107.

6 Brian Cronin, 'Comic book urban legends revealed #28', on *Comics Should Be Good* (8 December 2005), < http://goodcomics. blogspot.com/2005/12/comic-book-urban-legends-revealed-28. html>.

7 @hermanos (David Brothers), Twitter (26 November 2017).

8 Steven Weinberg, *Dreams of a Final Theory* (London: Vintage, 1994), p. 120.

9 Weinberg, *Dreams of a Final Theory*, p. 120.

10 Hawking and Mlodinow, *The Grand Design*, p. 109.

11 Hawking and Mlodinow, *The Grand Design*, p. 111, emphasis added.

12 Stephen Hawking, *A Brief History of Time* (London: Bantam, 2011), p. 113.

13 Hawking, *A Brief History of Time*, p. 118.

14 J. M. Bardeen, B. Carter and S. W. Hawking, 'The four laws of black hole mechanics', *Communications in Mathematical Physics* (1973) 31(2), pp. 161–70.

15 Hawking, *A Brief History of Time*, p. 119.

16 This is the title of his chapter on the topic in *A Brief History of Time*.

17 Stephen Hawking, 'Particle creation by black holes', *Communications in Mathematical Physics* (1975) 43(3), pp. 199–220.

18 Hawking, *A Brief History of Time*, p. 119.

19 Hawking, *A Brief History of Time*, from the Appendix.

20 Lee Smolin, *The Trouble with Physics* (London: Penguin, 2008), p. 91.

6 Part 2

1 Rafael Bombelli, *L'Algebra* (1572), quoted in Paul Nahin, *An Imaginary Tale* (Princeton, NJ: Princeton University Press, 1998), p. 19.

2 Stephen Hawking, *A Brief History of Time* (London: Bantam, 2011), p. 69.

3 Hawking, *A Brief History of Time*, p. 128.

4 Hawking, *A Brief History of Time*, p. 128.

5 J. B. Hartle and S. W. Hawking, 'Wave function of the universe', *Physical Review D* (1983) 28(12), p. 2960.

6 Hartle and Hawking, 'Wave function of the universe', emphasis added.

7 Nahin, *An Imaginary Tale*, p. 104.

8 Hawking, *A Brief History of Time*, p. 153.

9 Hawking, *A Brief History of Time*, p. 153.

10 Personal communication.

11 Personal communication.

12 Stephen Hawking and Leonard Mlodinow, *The Grand Design* (London: Bantam, 2010), p. 135.

13 Hawking and Mlodinow, *The Grand Design*, p. 135.

7

1 Mariusz Pudzianowski, 'Mariusz Pudzianowski Mixed Martial Arts (MMA) career', <www.mariuszpudzianowski.net/mariusz-pudzianowski-mixed-martial-arts-mma-career>.

2 *20/20*, ABC Television (March 1998).

3 Stephen Hawking, *A Brief History of Time* (London: Bantam, 2011), p. 209, emphasis added.

4 Michael White and John R. Gribbin, *Stephen Hawking: A life in science* (London: Penguin, 1998), p. 284.

5 Hawking, *A Brief History of Time*, p. 208.

6 Hawking, *A Brief History of Time*, p. 131.

7 Frank Tipler, 'The mind of God', *Times Higher Educational Supplement* (1988) 832, p. 23.

8 J. D. Flynn, 'Catholics question Hawking's comments on John Paul II', Catholic News Agency (19 June 2006).

9 John Polkinghorne, 'Where God meets physics', Research, University of Cambridge (28 November 2011), <www.cam.ac.uk/research/discussion/where-god-meets-physics>, emphasis added.

10 Polkinghorne, 'Where God meets physics'.

11 Hawking, *A Brief History of Time*, p. 209.

12 Stephen Hawking and Leonard Mlodinow, *The Grand Design* (London: Bantam, 2010), p. 5, emphasis added.

13 John C. Lennox, *God and Stephen Hawking: Whose design is it anyway?* (Oxford: Lion, 2011), p. 18, emphasis added.

14 Albert Einstein, 'Physics and reality', *Journal of the Franklin Institute* (1936) 221, pp. 349–82.

15 Johannes Kepler, *Optics*, tr. William H. Donahue (Santa Fe, NM: Green Lion Press, 2000), p. 15, emphasis added.

16 Immanuel Kant, *Critique of Pure Reason*, tr. J. M. D. Meiklejohn (London: Bell & Daldy, 1871), 9, p. 35, <https://babel.hathitrust.org/cgi/pt?id=hvd.hn1tu1&view=1up&seq=13>.

17 Kant, *Critique of Pure Reason*, 9, p. 38, emphasis added.

18 Hawking and Mlodinow, *The Grand Design*, p. 7.

19 Hawking and Mlodinow, *The Grand Design*, pp. 45, 46, emphasis added.

20 Hawking and Mlodinow, *The Grand Design*, pp. 45–6.

21 Aage Petersen, 'The philosophy of Niels Bohr', *Bulletin of the Atomic Scientists* (1963) 19(7), p. 12, emphasis added.

22 Hawking and Mlodinow, *The Grand Design*, p. 45.

23 Hawking and Mlodinow, *The Grand Design*, p. 42.

24 Hawking and Mlodinow, *The Grand Design*, p. 47, emphasis added.

25 Hawking and Mlodinow, *The Grand Design*, pp. 41, 42

26 Hawking and Mlodinow, *The Grand Design*, p. 164.

27 Hawking and Mlodinow, *The Grand Design*, p. 136.

28 Hawking and Mlodinow, *The Grand Design*, p. 164.

29 Hawking and Mlodinow, *The Grand Design*, p. 139, emphasis added.

30 Hawking and Mlodinow, *The Grand Design*, p. 139.

31 Hawking and Mlodinow, *The Grand Design*, p. 140.

32 Hawking, *A Brief History of Time*, p. 206.

33 Hawking and Mlodinow, *The Grand Design*, p. 15.

34 Hawking, *A Brief History of Time*, p. 160.

35 Hawking, *A Brief History of Time*, p. 160.

36 Hawking, *A Brief History of Time*, pp. 160, 161.

37 Hawking and Mlodinow, *The Grand Design*, p. 172.

8

1 Proverbs 18.17 (ESV).

2 Arvind Borde, Alan Harvey Guth and Alexander Vilenkin, 'Inflationary spacetimes are not past-complete', *Physical Review Letters* (2003) 90, p. 151301.

3 Aron Wall, 'Did the universe begin? II: Singularity theorems', *Undivided Looking* (25 May 2014), <www.wall.org/~aron/blog/did-the-universe-begin-ii-singularity-theorems>.

4 Stephen Hawking, *Brief Answers to the Big Questions* (London: John Murray, 2018).

5 John Matson, 'Artificial event horizon emits laboratory analogue to theoretical black hole radiation', *Scientific American* (1 October 2010), <www.scientificamerican.com/article/hawking-radiation>.

6 The authors are grateful for Tom Lancaster's personal correspondence.

7 See A. Vilenkin, 'Quantum origin of the universe', *Nuclear Physics* (1985) B252, pp. 141–51, or R. Bousso and J. Polchinski, 'Quantization of four-form fluxes and dynamical neutralization of the cosmological constant', *Journal of High Energy Physics* (2000) 06, p. 006 for examples.

8 Robert D. Klauber, 'Simplified guide to de Sitter and anti-de Sitter spaces', <www.quantumfieldtheory.info/dS_and_AdS_spaces>.

9 Don N. Page, 'Susskind's challenge to the Hartle–Hawking no-boundary proposal and possible resolutions', <https://arxiv.org/abs/hep-th/0610199>.

10 Page, 'Susskind's challenge to the Hartle-Hawking no-boundary proposal and possible resolutions'.

11 Job Feldbrugge, Jean-Luc Lehners and Neil Turok, 'No rescue for the no boundary proposal', <https://arxiv.org/abs/1708.05104>.

12 Quentin Smith, 'Two ways to prove atheism' (1996). Full text of his speech can be found at: <https://infidels.org/library/modern/quentin_smith/atheism.html>.

13 Aron Wall, 'Fuzzing into existence', *Undivided Looking* (22 July 2014), <www.wall.org/~aron/blog/fuzzing-into-existence>.

14 Reed Guy and Robert Deltete, 'Emerging from imaginary time', *Synthese* (August 1996) 108(2), pp. 185–203.

15 Aron Wall, 'Did the universe begin? IV: Quantum eternity the-
    orem', *Undivided Looking* (31 May 2014), <www.wall.org/~aron/
    blog/did-the-universe-begin-iv-quantum-eternity-theorem>.

16 William Lane Craig, 'Stephen Hawking: the anti-philosophy
    philosopher', *Reasonable Faith* (7 February 2011), podcast at:
    <https://youtu.be/5nyr6i0ppxY>.

17 Stephen Hawking and Leonard Mlodinow, *The Grand Design*
    (London: Bantam, 2010), p. 50.

18 Hawking and Mlodinow, *The Grand Design*, p. 50.

19 Hawking and Mlodinow, *The Grand Design*, p. 51, emphasis
    added.

20 Hawking and Mlodinow, *The Grand Design*, p. 172, emphasis added.

21 Craig, 'Stephen Hawking: the anti-philosophy philosopher'.

22 Personal correspondence.

23 Personal correspondence.

24 John Leslie, *Universes* (London: Routledge, 1996), p. 66.

25 John Lennox, *God and Stephen Hawking* (Oxford: Lion Hudson,
    2010), p. 49, emphasis added.

26 Don. N. Page, 'Does God so love the multiverse?' (2008), <https://
    arxiv.org/pdf/0801.0246.pdf>.

27 Wall, 'Fuzzing into existence'.

28 Luke A. Barnes, 'The fine-tuning of the universe for
    intelligent life', *Publications of the Astronomical Society
    of Australia* (2011) 29(4), p. 529, <https://arxiv.org/
    pdf/1112.4647.pdf>.

29 Lee Smolin, *Time Reborn* (London: Penguin, 2014), p. 131.

30 Lewis Campbell and William Garnett, *The Life of James Clerk
    Maxwell* (London: Macmillan, 1882), p. 158.

31 Hebrews 1.1–3 (NIV), emphasis added.

32 Hawking and Mlodinow, *The Grand Design*, p. 51.

33 Hebrews 1.1–2 (NIV), emphasis added.

34 Don N. Page, 'Hawking's timely story', *Nature* (April 1988), 332,
    pp. 742–3.

35 Page, 'Does God so love the multiverse?'.

36 Matthew 16.15 (Berean Study Bible, 2016), emphasis added.

Appendix 1 A note on M-theory

1 Stephen Hawking and Leonard Mlodinow, *The Grand Design* (London: Bantam, 2010), p. 180.

2 Lee Smolin, *The Trouble with Physics* (London: Penguin 2008), p. xiv.

3 As quoted in John Lennox, *God and Stephen Hawking* (Oxford: Lion, 2010). Original broadcast was 'Unbelievable?' with Justin Brierley (25 September 2010).

Appendix 2 A universe from nothing

1 Peter Atkins, *Conjuring the Universe* (Oxford: Oxford University Press, 2018), p. 18.

2 Lawrence Krauss, *A Universe from Nothing* (London: Simon & Schuster, 2012), p. 149.

3 David Albert, 'On the origin of everything', *New York Times* (23 March 2012).

4 Luke Barnes, 'A universe from nothing? What you should know before you hear the Krauss–Craig debate', Letters to *Nature* (13 August 2013).

5 Stephen Hawking and Leonard Mlodinow, *The Grand Design* (London: Bantam, 2010), p. 180.

6 Krauss, *A Universe from Nothing*, p. 149.

Acknowledgements

Excerpts from *A Brief History of Time: From Big Bang to black holes* by Stephen Hawking, published by Bantam, reprinted by permission of The Random House Group Limited. © 2011.

Excerpts from *A Brief History of Time: And other essays* by Stephen Hawking, copyright © 1988, 1996 and 2011. Used by permission of Bantam Books, an imprint of Random House, a division of Penguin Random House LLC. All rights reserved.

Excerpt from *The Collected Stories* by Arthur C, Clarke, copyright © 2001. Used by permission of Gollanz. All rights reserved.

Excerpts from *The Grand Design* by Stephen Hawking and Leonard Mlodninow, published by Bantam, reprinted by permission of The Random House Group Limited. © 2011.

Excerpts from *The Grand Design* by Stephen Hawking and Leonard Mlodninow, copyright © 2010. Used by permission of Bantam Books, an imprint of Random House, a division of Penguin Random House LLC. All rights reserved.

Excerpts by C. S. Lewis © copyright CS Lewis Pte Ltd.

Every effort has been made to seek permission to use copyright material reproduced in this book. The publisher apologizes for those cases in which permission might not have been sought and, if notified, will formally seek permission at the earliest opportunity.

Index

Index

Index

Index